Animal
Migration

Animal Migration

John Cloudsley-Thompson

BOOK CLUB ASSOCIATES ~ London

Contents

Introduction

A galloping herd of wildebeest, a flock of birds on the wing, or a salmon battling through white water are all stirring reminders of the fact that animals have great mobility. This mobility is a major factor in determining the lifestyle of many species. Any animal may be beset by problems of food shortage, overcrowding, predation or habitat destruction, and very often the best solution is simply to move to 'greener pastures'. The success and diversity of animals depend on flexibility and versatility, and it is hardly surprising that, with a few important exceptions, natural selection has favoured the evolution of mobile species.

There are three basic types of animal movement: trivial, nomadic and migratory. Trivial movement is found mainly among lower animals which are usually of fairly sedentary habit and move only within a limited area. Thus the Common Limpet (*Patella vulgata*) may, when it is browsing on the

Facing page: king penguins share their Antarctic beach with other marine animals. Smaller than emperors, king penguins rear their chicks during the short southern summer and the harsh winter that follows; right: when they are covered with sea water, limpets move around grazing on algae. Before they are exposed by the retreating tide, however, they invariably return to the same position on the rock

seashore, wander as much as a metre from its scar on the rock face; but it never fails to return to its 'home' when the tide recedes, because its shell has grown to fit a particular area of the rock surface.

Nomadism is the practice of leading a wandering life and is based on the need to find food. It is characterized by the regularity of the movements involved. This is particularly apparent in the case of animals, such as the moose (*Alces alces*), which make a restricted area their home for part of the year and are only truly nomadic at other times. In winter, moose collect together and retire to some sheltered area where they remain until the spring returns. During the warmer seasons, mooses are nomadic over a vast expanse of country.

The third type of movement, migration, is a regular to-and-fro directed movement with a complete cycle. It covers a wide variety of behaviour. Some migratory birds, for instance, breed within the Arctic Circle and winter south of the Equator. Many other species move only a few hundred kilometres. Some migrations take place several times within the life of an individual, others are undertaken only once. The Common Eel (*Anguilla anguilla*) swims up rivers as an 'elver', in its youth. Years afterwards it returns to the sea, there to breed and die. Other fishes come and go, year after year.

Movement of animals, such as lemmings, crossbills and waxwings, from their home territory is known as 'emigration'. In the past, this was thought to differ from true migration because these movements are few and far between. However, emigration is now seen in its true light, as the initial stage of an irruptive (that is, irregular) migratory movement. Conversely, when migrants arrive in a new habitat, they are called 'immigrants' while, if they are moving all the time, they are said to be 'nomadic'. Although these terms are useful, they are not precise. Clear differences between periodic migrations, sporadic changes of abode, trivial movements and other fluctuations do not exist, and it must be remembered above all that such movements are merely part of the cycle of propagation and survival adopted by a particular species.

7

Origins of migration

The fundamental impulses of all dispersal, wandering and migration, are derived from the requirements of food, reproduction, and competition for space in which to live. When a herd of buffalo or wildebeest, for example, doubles in number, its members have to roam over a much larger area than before in order to obtain sufficient food. They tend to follow the grass and, where this is seasonal, the beasts make seasonal migrations. Before the settlement of North America, the American Bison (*Bison bison*) used to move twice annually, between Canada and Mexico.

The advantages of migration are easy to understand. The origin of the phenomenon is, however, almost entirely a matter for speculation. It is, of course, bound up with the evolution of the species that display it. Periodic migrations cannot have come about suddenly. They are the cumulative result of the movements of countless generations. Animals that went the wrong way would have perished. Those that took the correct direction, on the other hand, would have survived and returned with their progeny. At first they need not have covered any great distances—just enough to find unoccupied ground—but the annual repetition would have become an established habit that eventually developed into a basic instinct within the population.

The evolutionary significance of migration has often been disputed. It is sometimes an expensive process for a species, since a large number of individuals may, as in the case of the lemming, fail to find a suitable new habitat and, consequently, die without leaving any offspring. At the same time, the migratory instinct clearly derives from natural selection. It is, at first sight, quite difficult to conceive of a selective mechanism operating in favour of such a lethal character. We are, therefore, forced to conclude either that when irruptions occur, more of the migrants survive than appears to be the case, or else possibly to accept the highly controversial suggestion that 'group selection' may be operating to some extent. Thus a group would benefit from the loss of some of its members because there would be fewer individuals to compete with one another when food was scarce.

Can such vast, unsuccessful, irruptions as those of Pallas's Sandgrouse (*Syrrhaptes paradoxus*) from the steppes and deserts of Central Asia, far into Europe, be explained by group selection? Most biologists think not, preferring a simpler explanation—that it is advantageous for these birds, and especially for the juveniles, to emigrate in times of food shortage. If they were to stay to compete with experienced adults they would almost certainly perish; whereas, by migrating, they have some small chance of surviving elsewhere and of returning afterwards to their natural home. For the moment, the evolu-

tionary significance of migration remains one of science's many unsolved mysteries.

Migration may have evolved gradually in the face of slow climatic changes, such as the retreat of the ice ages, which had driven birds and bats from their original homes. As the ice melted, a gradually increasing area of land became suitable for feeding and reproduction. This seems a more likely reason for migration than the suggestion that, through millions of years, a species could have retained an urge to return to its former home.

Some authors have suggested that certain migration routes reveal the geographical conditions of earlier epochs. For instance, they claim that the numerous birds which cross the Sahara at its widest part are following the route used by their remote ancestors when the desert was greener and more fertile, and that birds migrating from India to Madagascar are flying across a hypothetical land-bridge that once linked India with Africa. The theory of Continental Drift has also been enlisted to explain some of the problems of the origins of bird migration. It has been suggested that migrations between breeding and feeding areas gradually became longer as the continents drifted apart from one another. Alternatively, migration may have arisen suddenly, following irruptions like those of Pallas's Sandgrouse.

None of these ideas are necessarily incompatible with one another. Migration may have arisen as a

Right: a Neolothic artist has recorded his impression of game in prehistoric times when migration must have been a common sight. This painting of bison adorns the walls of the caves at Altamira; below: sanderling. These and other small waders breed in high northern latitudes from where they perform long migrations. When not breeding, they inhabit the shore and are highly gregarious

combination of climatic changes and range expansion caused by many different factors. In the case of trans-equatorial migration, where the two seasonal areas lie far apart, it is probable that numerous complex factors will have been involved. Each hypothesis must, however, remain in the realm of conjecture, since it cannot be substantiated by observation or tested by experiment.

Early theories

Over the centuries, conjecture has played an important, and often misleading, part in the theories about migration. These curious periodic movements have interested mankind since the days when primitive hunters first began to follow migrating herds across the wide savannas of tropical Africa. On rocks and cave walls—at Lascaux, Altamira and Tassili—magico-religious masterpieces in vivid colour portray the horses, bison and aurochs (extinct wild oxen) on which, for millennia, our ancestors depended for food and sustenance.

Long before Biblical times, the peoples of Persia and Arabia compiled portions of their calendars from the times of arrival and departure of certain bird species. Those whose appearance heralded warmer weather were greeted by festivals in their honour.

One early recorded—and frightening—migration was the eighth plague of Pharaoh, a plague of locusts of such vast numbers and remorseless progress that it darkened the land of Egypt for three whole days. 'Like the noise of chariots on the tops of mountains shall they leap,' wrote Joel (*ii*. 3–10):

They shall run upon the wall, they shall climb up upon the houses; they shall enter in at the windows like a thief. The earth shall quake before them; the heavens shall tremble; the sun and the moon shall be dark, and the stars shall withdraw their shining.

Rather more welcome was the flight of quail described in the Book of Numbers (*xi*. 31). Blown off course during their annual migration, they saved the Israelites from starvation in the wilderness.

In many of the Old Testament references to migration is the implicit and vital understanding that migration is not an odd, unexpected aberration, but a natural part of the cycles of life. In Jeremiah (*viii*. 7), for instance:

The stork in the heaven knoweth her appointed time; and the turtle-dove, and the crane, and the swallow, observe the time of their coming.

Even after the settlement of agricultural com-

9

Grey Whale

Monarch Butterfly

Monarch Butterfly

Pacific Salmon

Pacific Salmon

Arctic Tern

Arctic Tern

Eel

Eel

Barn Swallow

Abdim's S

Green Turtle

Bobolink

Bobolink

Green Turtle

Arctic Skua

A selection of routes of some outstanding migrants. Migration takes place on land, sea and air, and often requires phenomenal endurance

Arctic Tern

Arctic Tern

Wandering Albatross

Arctic Tern

Ruff

Ruff

Ruff

Grey Whale

Barn Swallow

Abdim's
Stork

Lesser Cuckoo

Spine-tailed Swift

Spine-tailed Swift

Wandering Albatross

munities, man must have wondered why certain fishes, birds and mammals were only present for part of the year, mysteriously disappearing at a certain season, only to reappear, equally mysteriously and regularly, several months later.

Despite such observations and speculations, no serious study of bird migration took place before the time of Aristotle (c.384–322 BC). His theories in the eighth book of *Historia Animalium* influenced belief on the subject for many centuries. After pointing out that some creatures can make provision against change of season without stirring from their normal haunts, Aristotle added that others migrate:

> quitting Pontus and the cold countries after the autumnal equinox to avoid approaching winter, and after the spring equinox migrating from warm lands to cool lands to avoid the coming heat. In some cases, they migrate from the ends of the world, as in the case of the Crane for these birds migrate from the steppes of Scythia to the marshlands south of Egypt, where the Nile has its source.

Enumerating many of the birds which were migratory in their habits—including pelicans, that 'migrate and fly from Strymon to the Istar and breed on the banks of the river'—he added:

> All creatures are fatter in migrating from cold to heat than in migrating from heat to cold; thus the quail is fatter when he emigrates in in autumn than when he arrives in spring. The migration from cold countries is contemporaneous with the close of the hot season.

The most provocative views on migration propounded by this great thinker were those on hibernation and the part it played in the life of birds. He believed that hibernation might account for the periodic appearance and disappearance of certain birds, including swallows, kites, storks, larks, thrushes, starlings, owls, turtle-doves and other 'cushats' (doves and pigeons). This erroneous theory held sway for many centuries after he first advanced it.

Aristotle also proposed a fanciful hypothesis of 'transmutation', to account for the seasonal emigration of some species of birds and the simultaneous appearances of others. He asserted that the European robin (*Erithacus rubicula*) and redstart (*Phoenicurus phoenicurus*) changed into one another, as did the blackcap (*Sylvia atricapilla*) and garden warbler (*S. borin*). He even claimed that these birds had actually been observed in the act of changing. This myth has persisted until modern times in some country districts of Britain where it is still said that cuckoos change into hawks in late summer and re-assume their normal appearance in spring.

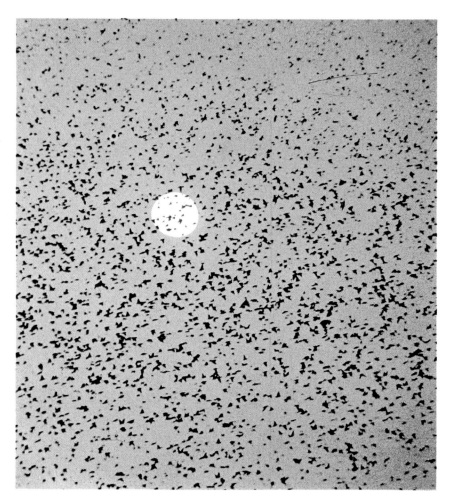

In 1555, Olaus Magnus, Archbishop of Uppsala in Sweden, published a curious work, *Historia de Gentibus Septentrionalibus et Natura*, in which he alluded to the seasonal disappearance of swallows and evolved the astounding hypothesis that they actually descended into pools of stagnant water and spent the winter months there. This absurd idea, suggested perhaps by the swallows' habit of feeding low over standing water, where insects collect, persisted for many years. So did the equally improbable theory that geese grew from barnacles that fell into the water from trees at the edge of the sea.

That acute observer of human behaviour, Samuel Pepys (1633–1703), did not prove as astute on animal behaviour. He came up with the same ideas as Archbishop Magnus:

> Swallows are often brought up in nets out of the mudd from under water, hanging together to some twigg or other, dead in ropes, and brought to the fire will come to life.

Even Dr Samuel Johnson (1709–84) held the same opinion, though he did not go so far as to say that the birds were actually dead:

> Swallows certainly sleep all winter. A number of them conglobulate together by flying round

Above: migrating starlings; their seasonal disappearance was once thought to be due to hibernation; above, right: the rapid, powerful flight of the migrating brent goose, often seen in large flocks low over the water

and round and then, all in a heap, throw themselves under water and lie in the bed of a river.

Francis Willughby (1835–72), a noted English ornithologist, was more prosaic. He wrote that natural historians were uncertain what became of swallows in winter, whether they flew to other countries or became torpid in hollow trees and similar places. He himself believed it probable that they flew to hot countries such as Egypt and Ethiopia, and he suspected a similar pattern in the cuckoo. But the great Carolus Linnaeus (1707–78), the architect of modern classification, writing in his famous work *Systema Naturae*, thought that 'the chimney swallow together with the window swallow, demerges, and in spring emerges'.

Many contemporary writers also supported the doctrine of bird hibernation, and repeated each other's accounts of swallows being taken from the depths of lakes. Despite the experiments of the Italian naturalist Lazaro Spallanzani (1729–99), who found that swallows could not withstand temperatures much below freezing, hibernation and submersion were generally accepted as being correct explanations of the mystery. It was really not until the publication in 1789 of Gilbert White's classic *Natural History of Selborne* that hypothesis and speculation were replaced by empiricism and observation. But, although he undoubtedly supported the fundamental principles of bird migration, White still believed that swallows and martins actually hibernated and remained in Britain throughout the winter months. In Letter XI, we read that on 4 November 1771 House Martins (*Delichon urbica*) were gliding above the sea bank at Newhaven:

From this account, and from repeated accounts which I meet with, I am more and more induced to believe that many of the swallow kind do not depart from this island, but lay themselves up in holes and caverns.

Other, equally implausible, hypotheses have been proposed to explain migration—that birds retreated to the moon to spend the months of winter there, or that small birds were carried on the backs of larger species. By the end of the nineteenth century, however, the migration theory had come to be accepted as fact, and hibernation as fiction. Only recently have authentic examples been recorded of hibernation among birds.

For all these extraordinary tales, the truth, as far as it is now known, is in many ways even stranger than the imaginative fictions it has supplanted, as we shall see in the next chapter.

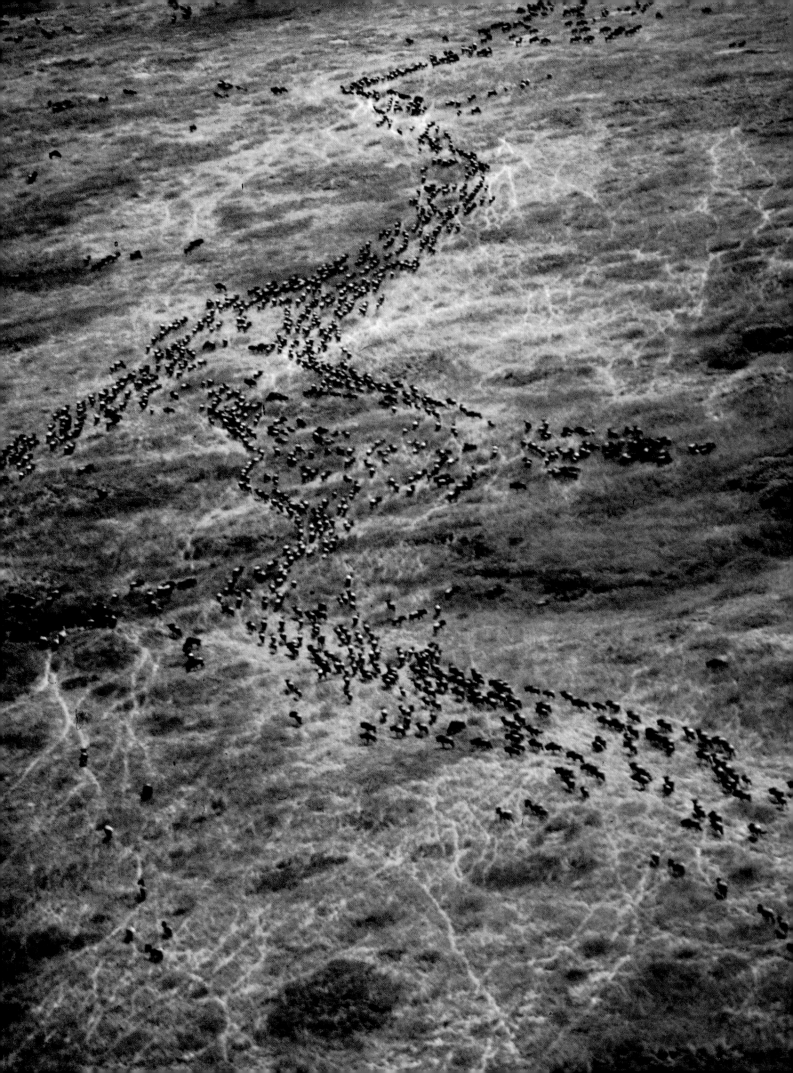

The Mechanics of Migration

If one were unexpectedly transported to a quite different part of the world, would it be possible, without a compass and without ever being told, to discover the direction of home? To some extent it probably would, and the unwilling traveller could do so—not through any mysterious sense of direction, but by the use of his or her eyes, memory, and sense of time. If his journey were northwards, and the flights had taken place in the northern hemisphere, the sun would be lower in the sky; whereas, had it been south, the sun would be higher. On the other hand, if his direction had been east, the time of day would immediately be later, just as in a westerly direction, it would become earlier. Across the Equator, the sun would appear to reverse its direction across the sky. One would really only need to remain long enough to determine in what direction the sun was moving before setting off on the long march homeward. It is by the instinctive use of such principles that migratory animals are able to navigate in the correct direction, and, as with our traveller, the essential requirements are memory, good eyesight, and an accurate sense of time.

For over fifty years it has been accepted that changes in day-length (photoperiod) are one of the most important factors governing the reproduction cycle in birds, as well as its associated activities, including migration. Differences in photoperiodic response mechanisms have led to a variety of annual breeding patterns, based on differences in the release of reproductive hormones. Nevertheless, the physiological stimulus for migration and reproduction is based on the interaction between a biological clock with an annual frequency, and changing photoperiod measured by means of the diurnal biological clock. Even in the case of many desert-birds, whose breeding occurs somewhat erratically at any season of the year, in response to the unpredictable and irregular rainfall, and whose secret of life is movement, reproduction seems to be not entirely free from the dictates of the photoperiodic clock.

Experiments have shown that, as in birds and other vertebrates, the annual sexual cycle of fishes is often regulated by photoperiodism. In species which migrate, that phenomenon, too, is often triggered by changes in photoperiod. Most amphibians undergo seasonal reproductive migrations in which not only visual and olfactory cues play a rôle, but gravity is also used for orientation in regions where the topography is abrupt and breeding pools lie in the bottoms of the valleys. In this connection, it is interesting that the thermal tolerance of frogs may also be affected by photoperiod. The tolerance of high temperatures is greatest during the late morning and early afternoon at the time of migration.

There is overwhelming evidence to show that migration and reproduction (and, in some cases, hibernation) in numerous species of mammals, as in birds, are largely controlled by day-length in temperate regions of the world. The time-sense required to measure photoperiodic changes is of the same order as that required for solar navigation, and day-length is likely to give a far more reliable indication of seasonal change throughout the year than alterations in any other climatic influence.

Navigation in birds

Every autumn, great flocks of starlings (*Sturnus vulgaris*) from southern Finland, Sweden and Denmark, fly through the Netherlands to winter in the south of Britain and the north of France. Migrating birds trapped in the country around the Hague, ringed, transported to Switzerland and released, behaved in one of two ways, as shown by the recapture of marked birds. All the young starlings, which had been hatched in the Baltic area and had not previously migrated, flew south-west on a course parallel to that normally taken. Older birds with previous migratory experience, however, tended to travel north-west into the region where they would have spent the winter had they not been disturbed. Since enough time had elapsed before their recapture to have permitted the birds to search a wide area, they might not necessarily have flown directly to their familiar winter territory. Nevertheless, the experiment does suggest that there are two kinds of orientation used. The simplest is 'directional' orientation which means that a particular compass

The Serengeti National Park, scene of mass seasonal migration

direction is selected. The second type is 'goal-directed' orientation.

The known migration routes of several species of birds, however, have been shown to follow Great Circles over the earth's surface. In such cases, this raises a major question about the validity of the map-and-compass model of bird navigation. Numerous facets of the navigation of birds still remain to be explained.

Like men, migrating birds have been shown to use both celestial and terrestrial navigation, according to the situations in which they find themselves. Thus, they may follow landmarks such as river valleys, mountain ranges, and the shoreline. The American hawks, which pass over the Isthmus of Panama twice yearly, deviate from their direct route to follow the western edge of the Sierra Madre. Similarly, gulls fly down the valley of the Werra in Hanover, and follow the Weser farther north until they reach the southern part of the North Sea. At other times, migrating birds steer by the sun, the moon, the stars, or perhaps by magnetic fields. Effective solar navigation, however, must make allowance for the apparent movement of the sun across the sky between dawn and dusk—in other words it must be time-compensated.

Numerous experiments in time-compensated solar navigation have been carried out on ringed birds transported to regions beyond their geographical range. In many cases, these have returned home so quickly that they could not possibly have spent much time in searching.

Examples include a Manx Shearwater (*Puffinus puffinus*) homing 4,800 kilometres (3,000 miles) to Wales from Boston, Massachusetts, in 12½ days, a Leach's Petrel (*Oceanodroma leucorhoa*) flying from Sussex, England, to Maine in 14 days, and a Laysan Albatross (*Diomedea immutabilis*) homing 5,150 kilometres (3,200 miles) to Midway Island from Washington State in 10 days. Releases from shorter distances have also resulted in homing times so brief that they could have been accomplished only by direct flight, and observations made *en route* have occasionally confirmed that the birds were on the direct line home. In all probability, the older starlings, in the experiment referred to above, would have flown directly to their correct winter quarters.

Many suggestions have been made to explain such phenomena, but it is now believed that the navigational process involves comparison of stimuli at the point of release with those remembered at home. Two hypotheses have been put forward. According to the first, the bird observes the movement of the sun along a small portion of its arc and then, by extrapolation, estimates the highest point and the time at which this would be reached. The sun's altitude at noon is inversely proportional to the observer's latitude, while longitudinal displacement is indicated by the difference between the time when

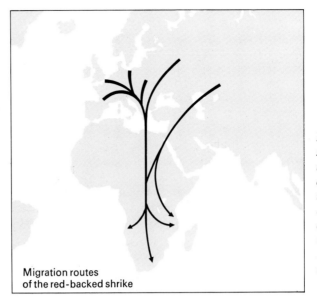

Migration routes
of the red-backed shrike

Below: the red-backed shrike, which breeds throughout most of central Europe including northern Spain, Italy, Greece and southern Sweden, is a summer visitor to Britain; left: the migration route

the sun is at its highest point in the sky and the time of noon at the bird's home. The second hypothesis suggests that the bird observes the slope of the segment of the sun's arc, and compares this with that which would be expected at the same time at home. Both methods of navigation would require very fine analysis of the sun's movement and the comparison of observed with remembered values.

The eyes of birds are peculiar in that they possess a curious, comb-like structure, known as the pecten, which projects over the 'blind spot' where the optic nerve enters the retina. Its function is completely unknown, but the suggestion has been made that the shadow it casts on to the retina may be used for measuring the sun's position. In other words, the pecten may perhaps operate like a nautical sextant, enabling birds to measure angles considerably more exactly than man can do without the help of optical instruments.

The method of navigation suggested by the first bi-coordinate hypothesis—that latitude and longitude are used to establish the birds' position—would require the apparent movement of the sun to be extrapolated, but reduces the memory requirement to a single point. The second method would necessitate memory of the whole of the sun's arc at the bird's home. Little of the experimental evidence upon which the first hypothesis was based has, however, been confirmed by subsequent workers. Furthermore, it is almost inconceivable that a bird should be able to measure changes in the horizontal movement or azimuth of the sun, while flying over the open ocean where there are no stable points of reference. In any case, each hypothesis requires a time-sense or 'biological clock' of extraordinary accuracy, but this is a faculty that birds, like many other animals, are known to possess. Of course, these calculations would be quite unconscious and the birds would instinctively turn in the correct direction. The results of experiments in which the biological clocks of birds had been advanced or retarded under artificial lighting before they were released away from home, indicate that they use the sun merely as a compass.

That the sun may indeed be used as a compass in bird navigation is confirmed by the fact that many species do not migrate either in fog, or in bad weather when the sun is not visible. (Birds that migrate at night probably steer by the stars, as we shall see.) In order to verify experimentally the hypothesis of time-compensated solar navigation, it would be necessary to alter the apparent position of the sun. This is obviously impossible with birds flying high in the sky, but information can be obtained in another way. It is known that warblers, starlings, and shrikes, among others, make what are called 'intention movements' at the time of migration. When perching, they sit with their heads

pointing in the direction in which they are going to migrate. Sometimes they flutter their wings, or even make short flights in that direction, before returning to their perches. They do this both in the wild and when kept in cages. When observed in experimental conditions, however, it was found that birds would only behave in this way when they could see the sun or the sky close to it.

Starlings placed in a cage, with a window through which they could see the sun, made migratory intention movements in the appropriate direction of their migration route. When a mirror was arranged outside the window, so that the direction from which the light came was altered by 90 degrees, the birds turned a similar amount from their original heading and steadfastly pointed in the new direction until the mirror was removed. Even birds hatched in an incubator and reared in a windowless laboratory with constant illumination, reacted in the same way when the season came for them to migrate, and they were allowed to see the sun. They would also respond equally well to artificial suns in the laboratory.

As a sequel to this, starlings have been trained to take food from a box in a particular compass direction from the centre of a six-windowed cage. Moreover, since they were able to pick out the correct box at any time of day, the birds must have been able to compensate for the sun's movement, and did not merely go to the box which was at a certain angle to the sun, whatever the time of day. Indeed, it was very difficult to teach them always to pick the box which lay in the direction of the sun; and they seemed to be quite incapable of learning to choose boxes in particular compass directions when they were in a room that gave no directional clues. This indicates very strongly that birds are able to orientate themselves by means of the sun.

The experiments described above were first carried out in Germany over 25 years ago, but subsequent research has served only to strengthen the hypotheses that they engendered. Homing Pigeons (*Columba livia*) and other birds have demonstrated the ability to return to their nests when taken away, either anaesthetized or in revolving, darkened, cages, so that they could not possibly memorize the route. Physical factors of the earth's rotation can largely be discounted, and the evidence so far unearthed points in the direction of time-compensated solar navigation being the most important method employed by migrating birds. The horizontal movement of the sun is allowed for: its altitude and vertical movements are ignored. It has recently been discovered that intention movements are performed most intensively by species that migrate for longer distances. This suggests that the distance to be flown is known instinctively while the direction is determined by the sun. But the degree of accuracy of the sun compass has not yet been finally answered.

Below: the indigo bunting of North America has been used extensively in experiments on bird navigation. When placed in a planetarium it obtains directional information from the pattern of the constellations near the Plough; facing page: migrating weaver birds, so-called because they build nests by inter-lacing grass and other vegetation

Many small birds, such as warblers, fly by night and must, therefore, be using a method of navigation that does not depend upon the presence of the sun. They make use of the stars, and are disorientated on cloudy nights.

A few years after the discovery that intention movements and pre-migratory restlessness are solar-orientated, it was found that European warblers, raised from the egg in constant conditions and in isolation, showed directed intention movements, not only under the natural sky, but in a planetarium in Germany. When the dome of the planetarium was illuminated with diffuse light, the birds would point randomly in different directions. When a blackcap was shown a simulated spring sky, however, it pointed to the north-east, as it might have been expected to do in natural conditions. Under a simulated autumn sky, on the other hand, it pointed more often to the south-west. A Lesser Whitethroat (*Sylvia curruca*), which would have been expected to

point south-east towards the Balkans, did just that more frequently than it pointed in other directions.

The time relation of apparent solar movements due to the rotation of the earth and the difference between the lengths of solar and stellar days, is extremely complicated. It would, therefore, be somewhat surprising if a time-compensation element were found to be involved in stellar navigation. To try and discover whether this element does in fact operate, an experiment was conducted in a planetarium in which Indigo Buntings (*Passerina cyanea*) were presented with skies appropriate to three, six, or twelve hours earlier or later. The birds' orientation was not deflected. It seems likely, therefore, that birds obtain directional information from the patterns of the constellation, much as people in the northern hemisphere determine north from the pointers of the Plough, or Big Dipper. Further experiments suggest that only those circumpolar constellations close to the Pole Star, which never pass below the horizon, are essential to the birds. Others have been eliminated from an artificial sky in a planetarium without affecting orientation.

The orientation of Indigo Buntings to the pattern of the stars also seems to be determined by the internal state of the birds rather than by seasonal changes in the constellations.

Recent experiments, carried out in San Francisco, have shown that bobolinks (*Dolichonyx oryzivorus*) make directional choices during migration restlessness. These birds perform very long migrations from well south of the Equator into southern Canada, but are more variable than warblers in the direction of their intention movements. In experiments, they showed a tendency to alternate between southward headings in autumn and exactly the opposite direction. Even when these south and northward pointings were alternating in an inexplicable fashion, however, some degree of celestial orientation was evident since the birds seldom pointed towards the east.

Until more detailed and extensive experiments have been carried out, it is not possible to know for certain whether birds are able to distinguish anything more than the difference between north and south from inspection of the stars at night. Unexplained results have been obtained from experiments with mallard (*Anas platyrhynchos*) which appear to learn more from the stars than from the sun because, when navigating at night, they are not disorientated by experimental re-setting of their internal timing mechanisms as they are when steering by the sun in daylight.

Mallard have been observed to orientate their direction of flight at night when the stars are obscured by a thin layer of cloud, although the moon remained visible. It is therefore probable that lunar navigation may be used by these and other birds. Although most birds are apparently able to use the

sun alone for navigation during the day, and although, by day, only goal-directed orientation has been demonstrated among caged birds, two points of reference obviously provide more information than just one. When the moon is visible by day, it may therefore provide an additional indication of direction to migrating birds.

Until 1965, the use of a magnetic sense in bird navigation tended to be dismissed as highly improbable—partly because it is difficult to imagine what type of sense organ could be involved, and partly on account of experiments in which pigeons homed equally well when small magnets were attached to their heads. But, of course, in these early experiments the birds were not prevented from seeing the sun. The fact that pigeons fitted with contact lenses of frosted glass have still been able to find their way home suggests, however, that navigation by visual means can be supplemented in exceptional circumstances.

That birds may in fact be able to respond to the earth's magnetic field is also suggested by recent experiments in Germany on European Robins that were exhibiting migratory restlessness. Although unable to see either a natural sky or the artificial one of a planetarium, the birds still made intention movements in a south-westerly direction. This is the course they would normally follow on their migration to Spain. Evidently they must have sensed something, through the walls of the building, which indicated to them the correct compass direction. When placed in a steel chamber, which cut off the earth's magnetic field, however, they fluttered to the bars of the recording cage without showing the slightest tendency towards any particular direction.

Further experiments have suggested that birds must become accustomed to altered magnetic relations for quite a while before they will respond to them. Once the need for acclimatization was taken into account, it was found that robins can orientate their movements—even when the night sky is invisible. Like a sailor on the high seas they steer by magnetic compass when landmarks are obscured.

It is not impossible that a magnetic sense may explain the mysterious homing and navigating ability of other animals, such as Weddell Seals (*Leptonychotes weddelli*) swimming beneath the dense pack-ice of the Antarctic. Homing Pigeons, too, have been disorientated when taken from Berlin and released near the Kyffhäuser Mountains in North Thuringia, where owing to underground deposits of iron, the lines of force of terrestrial magnetism are not normal. More recently, experiments carried out in America have suggested that the terrestrial magnetic field can be used not only for compass orientation but also for navigation. Homing Pigeons were taken from Fort Monmouth, New Jersey, to New Hanover, Pennsylvania, 145 kilometres (90 miles) to the west. Here they were released and

followed in an aeroplane. Measurements of terrestrial magnetism had established that either of two routes home would have had equivalent magnetic characteristics—and both routes were followed. When taken north, however, to a place from which a 'blind alley' of terrestrial magnetism led away from home, the birds were misled and circled aimlessly.

Experiments on gulls and other birds, mice, snails, termites, cockchafers and planarian worms suggest that a magnetic sense may be widely dispersed in the animal kingdom. The extent to which it is employed in migration is as yet unknown.

It is possible, too, that birds, such as pigeons, may learn to associate specific odours, carried by wind, with the direction from which it blows. Although

A flight of snow geese in South Dakota on migration to their breeding grounds in northern Canada

20

several well-designed experiments would appear to support this hypothesis it is surprising that birds should not possess better developed olfactory systems. Chemical sprays at Ben Gurion airport near Tel Aviv, and in nearby rubbish tips where birds congregate, are said not only to have reduced the number of birds congregating there but may cause the first flock that lands to leave behind chemical signals of fear which repel other birds. The suggestion has also been made that vultures may locate the general area of carrion by means of odour cues before the exact spot is pinpointed by vision. This, however, is not supported by field observations of vultures descending on the bodies of dead animals long before putrifaction has begun.

Navigation in arthropods

Many insects possess a highly developed sense of time. The ability of honey-bees (*Apis* spp.) to return to a source of food at the same time each day has been known since the beginning of the present century, when it was observed that bees timed their visits to a field of buckwheat only in the morning while the blossoms were secreting nectar. (Buckwheat is a cereal used in the preparation of breakfast foods and as fodder for domesticated animals.) In experiments, bees have been trained to visit the same feeding place—not only once each day, but even up to nine different times—and also to distinguish between intervals only 20 minutes apart. The timing of their visits is not affected by temperature or weather, and training is not disturbed when the experiments are conducted under constant conditions in the laboratory. Indeed time perception appears to be innate, since bees have been trained which had hatched in a dark chamber and had never experienced the alternation of day and night. The time-sense is not based on a learning of intervals deviating from a 24-hour periodicity—such as intervals of 19 or 27 hours. It is noteworthy, too, that they always visit the right place at a given training time.

The time-sense of honey-bees enables them to return to a known source of food when pollen and nectar are most readily available. Moreover, if the bees are compelled to remain inside the hive for a couple of days because of bad weather they can still remember when a particular plant secretes its nectar.

Many other unrelated insects are also able to learn the hours at which different kinds of flowers offer their nectar and pollen. Furthermore, the time-sense can be used for time-compensated solar navigation by insects as it is by birds.

To an observer in the northern hemisphere facing south, the sun rises on the left and sets on the right; but to an observer in the southern hemisphere who must face northwards to see the sun, it will appear to rise on the right, and set on the left. This explains the results of an experiment carried out in Brazil, in the southern hemisphere. Foraging bees from a local strain of *Apis mellifera*, long established in the region, were fed in the evening on a dish in a particular compass direction and transferred overnight to a new locality quite unknown to them. During the following day, the majority of the bees were, at all hours, searching in the direction of their previous training. On the other hand, the offspring of queens recently imported in an inseminated condition from the northern hemisphere, after similar training, showed systematically false orientation on the day of observation, since they tended to make a reverse allowance for the sun movement—as though it were in the northern hemisphere. The change in direction of compensation shown by the local strain must have occurred during the years that have

Above: worker honey-bee collecting nectar from a flower, after which it will return directly to its hive, navigating by the sun or polarized light in the sky; left: the compound eye of an insect, in this case a fly, is composed of numerous units. For orientation the image of the sun need not fall onto the same units, but must fall on those which are in the same azimuth. Units facing in the same direction appear to be linked in their nervous circuits

elapsed since the first honey-bees were shipped to Brazil sometime after 1530.

It has been known for many years that ants will use the position of the sun to maintain a straight course in territory lacking prominent visual landmarks. Using an innate time-sense, they are also able to allow for the movement of the sun, but only during summer. In spring, they cannot compensate for solar movement, and if detained for a few hours in a darkened box and then released, will continue at the same angle to the sun as before and, therefore, in a different direction. As yet, it is unknown why compensation for the changing position of the sun has to be learned anew after the winter by ants, while bees allow for it instinctively.

Sun-compass orientation reaches its highest perfection among bees and ants which are constantly leaving their nests on long journeys of reconnaissance. At any given moment they are able to return directly to their homes, although experiments with artificial suns have shown that it makes no difference to them whether the sun's height above the horizon corresponds with the actual time of day or season of year. They are not disorientated if it is 25° higher or 40° lower in the sky than it should be.

For correct orientation, the image of the sun need not fall on the same units, or facets, of the insect's compound eyes, but it must fall on units which are the correct azimuth—the angular distance from the north or south point of the horizon to the point where the vertical circle meets the horizon. Thus in summer, when the sun rises steeply in the morning, it changes its azimuth only slightly, but at noon, when it moves almost parallel to the horizon, it reaches its highest rate of change of azimuth, decreasing towards evening. The sun's rate of change of azimuth does not therefore proceed regularly, and daily differences in angular velocity vary both with seasons and geographical latitude.

Recent field experiments have suggested that ants

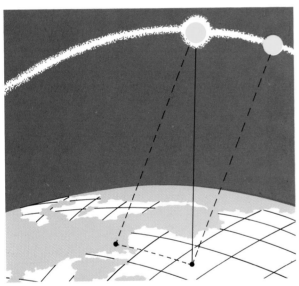

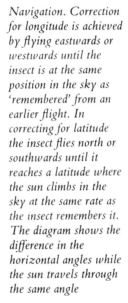

Navigation. Correction for longitude is achieved by flying eastwards or westwards until the insect is at the same position in the sky as 'remembered' from an earlier flight. In correcting for latitude the insect flies north or southwards until it reaches a latitude where the sun climbs in the sky at the same rate as the insect remembers it. The diagram shows the difference in the horizontal angles while the sun travels through the same angle

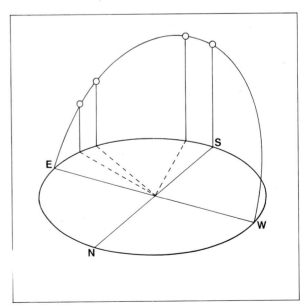

23

and bees can compensate for changes of solar azimuth with astonishing accuracy. This has been confirmed under experimental conditions with an artificial sun. Even when its course was altered so that it rose unnaturally in the west and set in the east, some ants quickly learned to use it as a means of orientation. As mentioned above, in bees such knowledge is inborn and is not relearned by individual insects.

Time-compensated solar orientation has also been observed in locusts, beetles, water-striders and wolf-spiders. Littoral sandhoppers, if moved inland and placed on dry soil, are able to return directly to the sea. This they do from observation of the sun. When specimens of the sandhopper *Talitrus saltator* were transported from the west coast of Italy to the Adriatic shore, they continued to move westwards on their release, despite the fact that the nearest sea now lay to the east.

The biological clocks of sandhoppers, like those of honey-bees and other insects, are very resistant to other environmental changes. When sandhoppers were taken by plane from Italy to South America, they orientated at an angle to the sun that would have been correct had they remained in Europe. The importance to a small crustacean, which would soon become desiccated in dry surroundings, of finding its way directly back to the damp sand of the seashore when accidentally transported inland or blown there by the wind, is quite obvious. Animals of each

Right: a South African wolf-spider with its young. Some species of wolf-spider have been shown to navigate by the sun; below: a sandhopper jumps across the beach. When blown inland, sandhoppers are able to find their way back to the shore by observing the direction of the sun or moon

population must, however, learn a different direction, according to the position of the sea in relation to the beach on which they live.

A similar type of orientation has been found in certain wolf-spiders which often live on the shores of lakes and rivers. When placed on the water, they hurry back to the bank, in a direction perpendicular to the shoreline. But, if taken to the opposite bank and placed on the water there, they attempt to cross to the shore on which they normally live, running in the direction to which they are adjusted by the position of the sun in the sky. That the angle towards the sun must be decisive, has been demonstrated by experiments in which the spiders were misled about the sun's position by means of a mirror. The participation of a sense of time has been confirmed by experiments in which wolf-spiders were kept temporarily in darkness and then released. Despite the movement of the sun overhead, the spiders headed directly for the home shore.

Insects not only navigate by the sun but are able to appreciate the plane of polarization of light and use this in their navigation. Light travels in waves which vibrate transversely—that is, at right angles to the direction of travel. In ordinary light these vibrations lie in an infinite number of transverse planes but, in polarized light, they are in only one transverse plane. The light reflected from any part of the blue sky is partly polarized, the plane and the degree of polarization depending on its direction in relation to that of the sun.

The human eye cannot distinguish between ordinary and polarized light, but insects and other arthropods can even distinguish the direction of the vibrations, a facility they make use of in their orientation. For this reason, honey-bees are in no way disorientated when the sun is obscured by cloud. So long as a part of the sky remains visible, they are able to maintain their sense of direction.

Although Italian populations of wolf-spiders are incapable of orientating themselves during the night, Finnish populations of the same species (*Arctosa perita*), living within the Arctic Circle, have been shown to orient themselves correctly throughout the

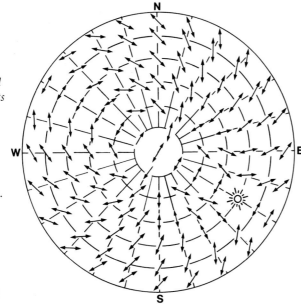

Above: cabbage white butterflies fly at an angle to the sun which varies according to the temperature experienced by the larvae. The insects shown here are mating

Right: polarization of sky light at about 10.00 hours, the double arrows showing the plane of polarization. The pattern alters with the course of the sun. Insects know these connections which they work out from the position of the sun even when it is clouded

24 hours on the days around the summer solstice. Between sunset and dawn, they direct their movements by the polarized light of the sky. During the Arctic summer, when there is continuous daylight, it is important for shore-dwelling animals to be able to orient themselves throughout the time they incur the risk of being blown into the water.

Study of butterfly migration in Britain has shown that the direction of flight is determined by temperatures experienced during larval development. Some species orient themselves by means of the sun, without compensating for its movement during the day, but others show no deviation from a constant flight direction. Cabbage Whites (*Pieris brassicae*) are among the species that do not show any time-compensation. However, the angle to the sun in which they fly varies according to the ambient temperature: in the spring, the first brood tends to fly north while the second brood—which matures in late summer—tends to fly south towards warmer temperatures. This change of direction in autumn is characteristic of species that overwinter as larvae (caterpillars) and is not found in species which hibernate in the encased, torpid pupal stage. The peak flight direction of butterflies that overwinter as immature adults is towards regions having a shorter winter rather than a milder one—for instance, from the south of England into Continental France.

Locusts also orientate themselves by the sun, but the direction in which they fly has little ecological significance. The function of the response is that it helps to keep the members of a swarm together since they fly in the same direction. As mentioned later in more detail, the net direction of movement of swarms is down wind into areas of low atmospheric pressure.

Little seems to be known about lunar navigation in arthropods, but it is not improbable that many nocturnal species should navigate by means of time-compensated reactions to the position of the moon. An insufficient quantity of food near the sea induces sandhoppers to travel towards the upper part of the beach at night when the humidity is high. At the same time, migration is inhibited by low humidity, and rainfall may elicit migration even during the hours of daylight. It is not clear whether the inland migration is directed, or merely represents casual wandering from the day-time habitat; but the return to the sea is certainly oriented either by the sun (as we have already seen) or by the moon. This is indicated by the fact that the animals are dis-oriented on moonless nights or when clouds obscure the moon, and their orientation is disturbed during the first and last quarter of the moon. If the moon is hidden, the animals will orientate themselves to the light of an electric torch, by forming an angle with the torch like that assumed with the moon, whatever the position of the torch.

Worker ants, marching in a straight line at night, are suddenly disoriented when the moon becomes obscured by clouds; and the assumption of a deter-mined angle of movement has been demonstrated by means of experiments with mirrors. This indicates that they, too, can steer by the moon as well as by the sun.

Navigation in fishes

Like insects, fishes are able to use the sun's position as a directional reference, and to remain oriented throughout the day by making allowance for the daily movement of the sun. This has been demonstrated by experiments in which various species of fishes have been trained to swim at certain angles to the sun at specific times. The angle of movement to the sun's position changes throughout the day, so the fishes must compensate for it. In other experiments, fishes were trained in the northern hemisphere and

then flown to the Equator and the southern hemisphere, where they continued to change their angle of movement relative to the sun's position during the day, just as they had in the original northern latitude.

Many fishes perform impressive migrations, both in the ocean and fresh water. Like marine turtles, which may swim for thousands of kilometres to the beaches on which they land and lay their eggs, they are probably able to do so in response to time-compensated solar navigation, assisted by their sense of smell. At the same time, marking experiments have shown that salmon and other fishes that live in the sea but breed in fresh water, return to the actual rivers in which they were born, when the reproduc-tion impulse overtakes them.

In 1880, F. Buckland suggested that 'when the salmon is coming in from the sea he smells about until he scents the water of his own river. This guides him in the right direction, and he has only to follow up the scent.' Recent research, too, suggests that directed migration leads Pacific salmon (*Oncorhynchus* spp.) from the high seas to the areas where their native rivers empty into the ocean. Here, olfactory cues are provided which the fishes presumably follow upstream to their spawning grounds. The evidence upon which this hypothesis has been based depends upon higher recovery rates of salmon tagged near to the coast than among those marked and released far out to sea. Whereas salmon presumably learn the smell of their birth-place, the response to fresh water of elvers, hatched in the Sargasso Sea, must be instinctive. Neverthe-less, it has been shown experimentally that young eels are not attracted to inland water that has been stored for a long while, or has been filtered through activated charcoal.

Navigation in mammals

Mammalian behaviour differs from that of birds and lower animals in that learning is far more important and instinct plays a lesser rôle. Consequently, it is not surprising to find far less evidence of celestial navi-gation among mammals, although few species have been investigated experimentally. Newspapers and magazines occasionally report on dogs or cats returning to their old homes after having been moved sometimes for several hundred kilometres, but no controlled experiments have yet been carried out on the subject.

The Striped Field Mouse (*Apodemus agrarius*), which is to some extent day active has, however, been shown to navigate by means of the sun. Quite possibly, the larger social mammals may learn their migration routes, when young, from their parents and other members of their species—and afterwards pass on this information to subsequent generations. Thus, the Grey Whale (*Eschrichtius gibbosus*) which migrates annually from its summer range in the Bering Sea to its breeding grounds in the warm,

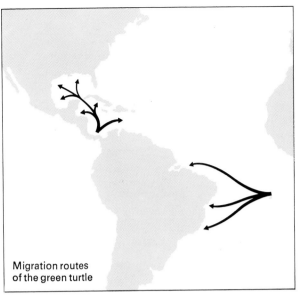

Migration routes of the green turtle

Above: female green turtle laying her eggs in the sand, after which she will find her way back to the seaweed pastures thousands of kilometres away; above, right: migration routes of the green turtle in the Atlantic from Ascension Island and Costa Rica where it breeds to feeding grounds off the coasts of Brazil and Central America

shallow, waters off the coast of California, may learn to follow the coastline.

Whales and porpoises probably steer by the sun and also follow the contours of the sea bottom by echo location; while bats, too, may be able to employ sonar for homing once they have learned their surroundings. Blindfolded bats are sometimes able to find their way home but, if their ears are plugged, they refuse to fly at all until their hearing has been restored. The assumption that the olfactory sense in mammals may also be involved has yet to be confirmed experimentally and we are still at the threshold of knowledge in this respect.

Navigation in amphibians and reptiles

Each spring, California Red-Bellied Newts (*Taricha rivularis*) descend from the wooded mountain slopes on which they live to selected streams where breeding takes place. Even if displaced for up to eight kilometres (5 miles) and released in some foreign river, a surprising number, over 60 per cent,

will return to their native stream to breed. Experiments have shown that olfactory cues play an important rôle in homing. Amphibian migrations seldom exceed a few kilometres, although those comparatively sluggish animals do tend to home to precisely defined goals. In the case of frogs and toads, navigation by the sense of smell may be assisted by sound and sight, and there is some evidence that time-compensated celestial orientation may also be used.

Sun-compass orientation, supplemented by odour gradients in the sea, probably explains the long-distance navigation of marine turtles. It has also been suggested that some odorous substances emanating from Ascension Island and travelling in the South Equatorial current westwards to the feeding grounds off the coast of Brazil may guide Green Turtles (*Chelone mydas*) back to the islands on which they breed. Recoveries of labelled bottles allowed to drift from the turtle breeding grounds of Costa Rica yielded a pattern similar to that of recoveries of marked turtles. On reaching maturity, it is not inconceivable that turtles could reverse the route down which they have drifted by the sense of smell, just as salmon do. Experimental attempts to demonstrate sun-compass orientation in terrestrial Box Turtles (*Terrapene carolina*) and lizards have produced interesting results and it is quite possible that time-compensated solar orientation occurs among reptiles as it does in so many other kinds of animals.

The ability to orientate is clearly of crucial importance to animals, and especially to migratory species. In many cases, visual landmarks are used, and time-compensated solar, lunar or celestial navigation is then unnecessary. But they are valuable stand-bys in cases of emergency and, when very long distances have to be traversed, as in the case of many birds, they often become the principle methods of navigation. One of the most absorbing aspects of animal navigation is that so much is yet to be explained.

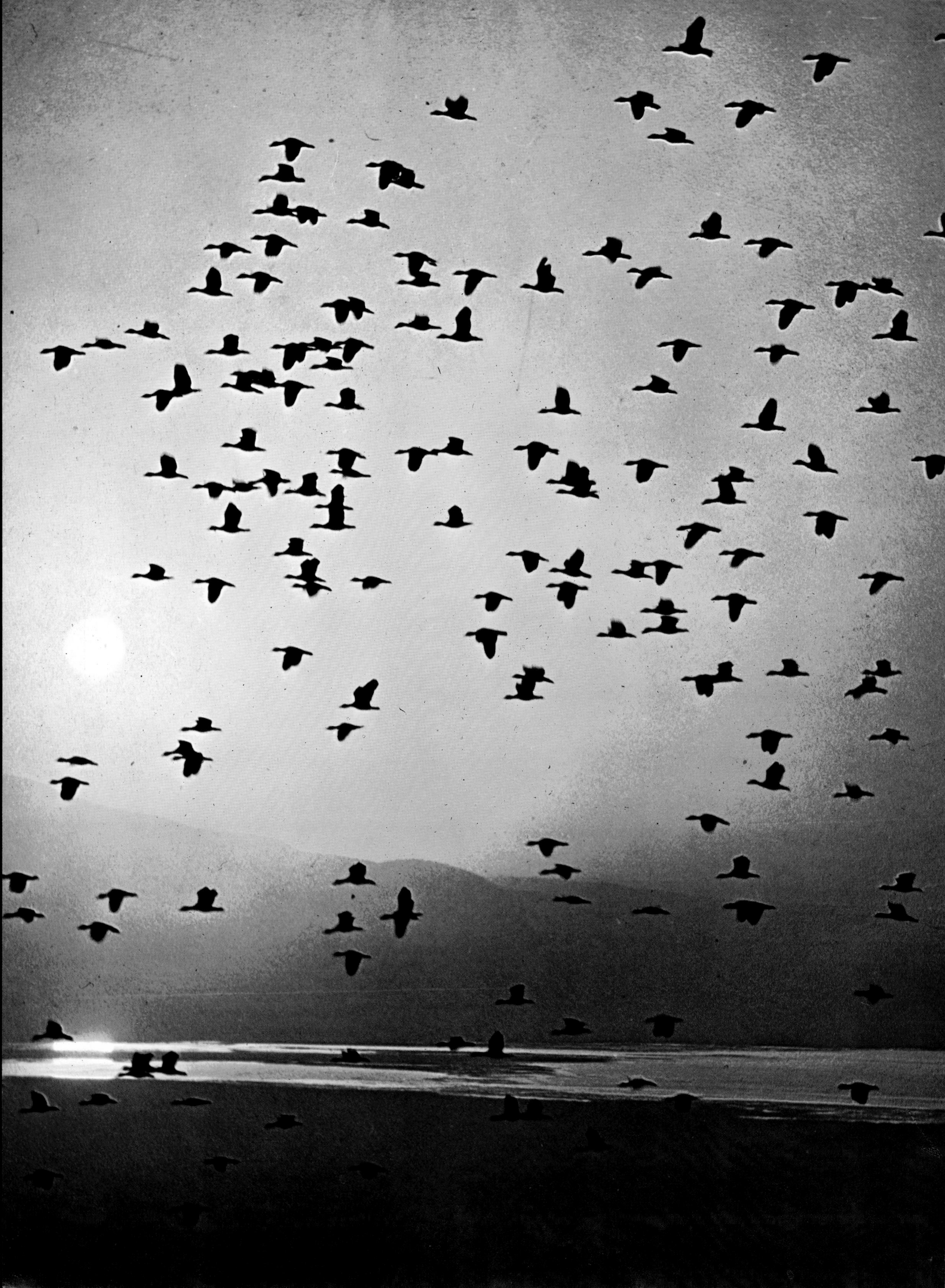

Migration by Air

Migration is such a distinctive feature of avian behaviour that birds are often classified according to their migratory status. 'Summer residents' are those that migrate northwards to spring breeding grounds, rear their young in summer and return to their wintering areas in the autumn. 'Winter residents' are those birds that move from their breeding grounds in the far north to spend the winter in a southern locality where food is more abundant, while 'transient visitors' are those birds that pass through a locality, usually twice each year—first in spring on their way to the breeding area, and again in the autumn *en route* for the winter habitat. Obviously, these descriptions depend on the geographic viewpoint of the observer—the summer residents of one locality would be the transient visitors to another. Finally, there are 'erratic wanderers' that have no permanent habitat except during the breeding seasons.

These categories of birds are represented widely throughout the northern hemisphere. Far fewer species, however, move southwards to South Africa or South America to breed there during the northern winter. The reason for this is simple—there is a much more extensive landmass in the northern

hemisphere than in the southern. Nevertheless, over 30 species of birds, including no less than nine kinds of cuckoo, are summer visitors to the southern temperate zone and breed in South Africa. Australia, too, receives some Asian birds as visitors from the north at that time of year and a few American birds, especially water-fowl, breed in the south and winter further north.

Flight adaptation in birds

Birds are the greatest and best-known migrants, able casually to exchange one man's winter for another man's summer, or so it has always seemed to those who wistfully watch their departure and eagerly await their arrival. There is, of course, nothing casual in these migration journeys, for each species is minutely adjusted to the lifestyle it has evolved.

The shape of a bird's wing is specially adapted to the type and range of its flight. The Wandering Albatross (*Diomedea exulans*), for example, has a wing adapted for dynamic soaring in strong winds. The bird glides down wind at great speed and then, close to the water, turns and gains height into the wind. Such a method of flight is only possible for a bird of considerable weight and size. The wing-span is about 3.35 metres (11 feet) but the wings are narrow and streamlined. In contrast, the wings of vultures are long and wide, engendering a low stalling speed which enables the birds to maintain height when gliding in weak upcurrents.

Of the birds that indulge in flapping flight, as opposed to gliding, the swifts are markedly adapted for high speeds. Their bodies are streamlined, and their wings are swept back and lack slots, while the flight muscles are extremely well developed. It is interesting to compare these birds with owls, whose wings are extremely large for their weight, enabling them to fly very slowly with a low flapping rate. The breast muscles are proportionally smaller than in any other group of birds, giving great economy in energy.

The four examples cited above—albatrosses, vultures, swifts and owls—represent the two extremes

Facing page: geese on migration, silhouetted against the autumn sunset

Right: global migration around Antarctica of the wandering albatross

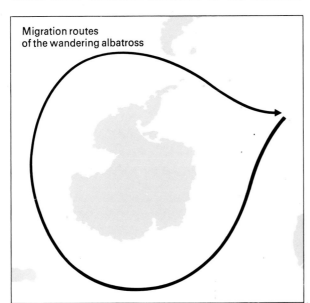

Migration routes
of the wandering albatross

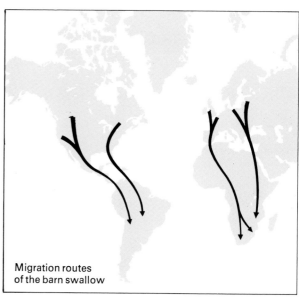

Migration routes
of the barn swallow

of gliding and flapping flight respectively. The types of flight found among other groups of birds come between these extremes, but are invariably adapted to their way of life.

The great bird migrants

Some of the best-known examples of migration are afforded by swallows, swifts and martins, because of their close association with man, and because, although they may winter thousands of miles away in South Africa, these small birds often return year after year to the same breeding places in Europe. Their arrival heralds the coming of the European spring, just as their departure forewarns the advent of winter.

Swifts, martins and swallows are rapid, powerful fliers, and do not hesitate to cross vast expanses of water or desert, although the flying insects on which they feed are rare in such places, especially in windy weather. The House Martin (*Delichon urbica*), for instance, crosses the Sahara at high altitude in a single, non-stop flight. These birds cover such vast distances because the long summer days of the northern temperate regions, accompanied as they

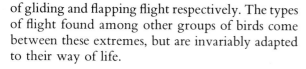

Above: the wing shapes of the swift and owl allow fast and slow rates of flight respectively; below: wing silhouettes of the vulture and albatross

are by huge increases in the numbers of insects available, provide both ample food for the newly hatched young and extensive hours of daylight during which the parents are able to feed them. The sooner that a chick is big enough to fly to its winter quarters in the south the better will be its chances of survival; so continual feeding and rapid growth are highly beneficial. Thus, like most irruptive emigration, breeding migration is ultimately associated with nutrition.

Swallows and martins will even use the same nest year after year if it has not been destroyed. Baby birds must suffer torture from the grim hordes of hungry parasites which inhabit their nests; and bird-lovers who preserve these habitations from one year to another unwittingly preserve the hibernating louse-flies, fleas, mites and bugs which are responsible for pain, disease and even death to the nestlings of the following spring.

The swifts, swallows and martins are all compelled to fly southwards before the approach of cold weather when their insect prey vanishes. Of the species breeding in Europe, the two most common are the Barn Swallow (*Hirundo rustica*) and the House Martin. Their autumn migration to Africa has been confirmed many times by the recovery of ringed birds. Swallows from Britain usually winter in South Africa, while birds ringed in Germany are more often recovered from Zaire, Nigeria, Morocco and Tunisia. Although it is impossible to assign a definite winter territory to each local European population, certain general tendencies have been established.

When unfavourable climatic conditions kill or disperse the insects upon which they depend for food, large numbers of swallows, martins and swifts may die from starvation. The appalling losses that swallows occasionally suffer may have given rise to the legend that they hibernate in winter. Small birds naturally tend to suffer heavier losses when on

migration than larger species, as they do at all times, but they compensate by laying more eggs.

Migration is by no means a race against time. Some swallows leave their summer homes at the end of July while others remain until the end of September, their departure depending upon the abundance or scarcity of their insect food. The birds move southward by day in large flocks, feeding as they fly. At sunset they rest, choosing marshy regions where reeds and rushes afford them shelter for the night. Both the autumn migration and the long return flight are accomplished over a period of five or six weeks, but the flocks are smaller in spring.

To cross the Sahara, House Martins use all their reserves of food and energy, but pelagic birds, such as petrels and albatrosses, can cover great distances with little exertion or apparent fatigue. From a few restricted nesting areas they are able to roam over a large part of the oceans of the world, gliding majestically on their long, narrow wings. Although albatrosses range so far over isolated regions of the oceans, they return regularly each year to particular breeding places.

On one of the Midway Islands, a breeding colony of Laysan Albatrosses was interfering with flights from a new airbase. In a trial experiment, therefore, 18 adult birds were taken by plane to places that, it was hoped, would become new homes, including

The barn swallow, top left, and house martin, top right, are compelled to fly south before the winter, when their insect prey disappears; above, centre, is the barn swallow's migration route; right: the black-crowned night herons who nest in Europe fly to tropical Africa for the winter, where they hunt the river banks for invertebrates, amphibians and reptiles as well as fish

Migration routes
of the greater shearwater

Tristan da Cunha

*Left: migration route
of the greater shearwater
from its breeding ground
on Tristan da Cunha;
right: emigration of
Wilson's storm petrel;
far right: Arctic tern
and its migration routes*

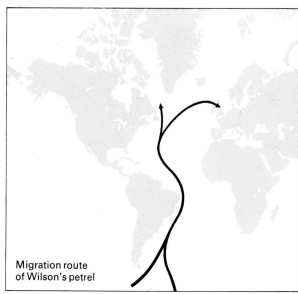

Migration route
of Wilson's petrel

Washington, Alaska, Japan, New Guinea and Samoa. Within a couple of weeks, 14 of the birds had returned to Midway—the fastest after only ten days. The staff of the airbase now always have to ensure that the runways are clear of albatrosses before planes can land or take off.

The Greater Shearwater (*Puffinus gravis*) nests from January to March on Tristan da Cunha. After the breeding season, it flies northwards on a circular route to Newfoundland, Greenland, Iceland and the Faroes. Then it swings south to the Azores, finally returning to Tristan da Cunha. The Sooty Shearwater (*P. griseus*), likewise, breeds in the southern hemisphere—in New Zealand—and spends the summer in Labrador, Greenland and Iceland. Wilson's Storm-Petrel (*Oceanites oceanicus*) breeds in Antarctica and then enjoys the summer season in the North Atlantic, crossing the ocean in a great loop. A comparison of the directions of the flights with those of the prevailing winds suggests that the

*Below: the sooty
shearwater breeds in
burrows on islands in
the southern hemisphere,
visiting the North
Atlantic in summer and
autumn; right: the
white stork is a
summer visitor to
Europe where it breeds.
The winter is spent in
Africa*

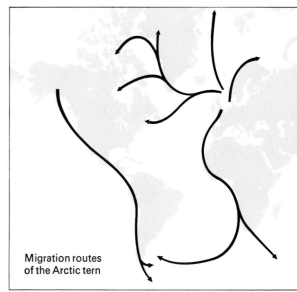

Migration routes
of the Arctic tern

petrels are borne northwards by the south-east trade winds which hurry them across the tropical seas where little food is available. Next, they are carried more slowly by westerly winds across the North Atlantic—where food is plentiful—and then south-east by the winds blowing from the African coast. Migratory sea-birds never attempt to struggle against strong winds.

Many of the world's sea-birds are long-range migrants, but few of them undertake journeys more extensive than those of the Arctic Tern (*Sterna paradisea*) or the Short-tailed Shearwater (*Puffinus tenuirostris*). The latter breeds on the shores of Tasmania, where it is known as the Mutton Bird, as well as in south-eastern Australia and on many of the islands of the southern Pacific. Its total numbers have been estimated in tens of millions. Short-tailed Shearwaters leave their nesting territories late in April, and fly out to sea in a huge circle which takes them round the eastern seaboard of Australia and northwards to Japan. This stage of 8,850 kilometres (5,500 miles) is sometimes covered in a month. June, July and August are spent in the Pacific Arctic before the birds return, 'riding' the trade winds and westerlies on their homeward journey across the eastern Pacific.

The White Stork (*Ciconia ciconia*) is another bird whose migrations have been extensively studied. Nesting along a broad strip from the Netherlands to western Russia, as well as in parts of Spain and N. Africa, the European stork population is divided into two distinct parts according to their migration routes. Birds that breed in western Europe fly through France, Spain, over the Straits of Gibraltar and down to West Africa; while those that reproduce in eastern Europe and Asia—the great majority—fly across the Bosphorus, through Turkey and Palestine down to East and South Africa. These separate routes are probably explained by the fact that storks will not fly across great expanses of water where the rising

33

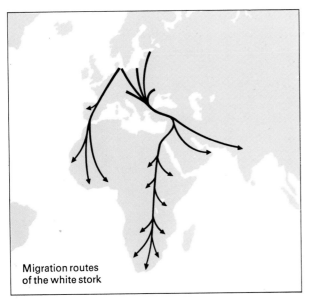

Migration routes
of the white stork

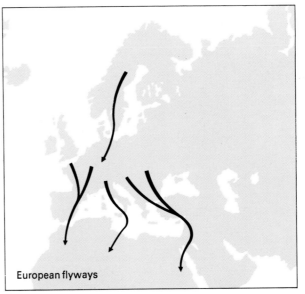

European flyways

*Far left: migration
route of the white stork;
left: European flyways;
below, left: white storks
soaring prior to their
autumn migration to
Africa; below:
migration route of the
willow warbler; right:
typical formation of
migrating common
crane; below: the
crane is a
nocturnal as well as
diurnal migrant*

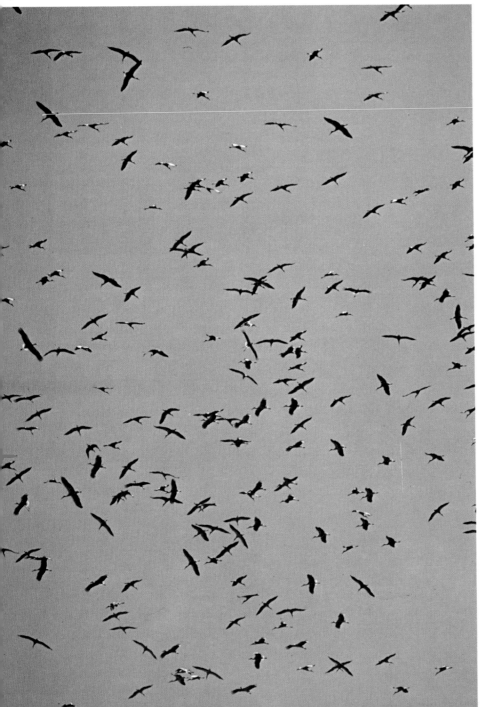

air currents on which they glide are absent. Migrating flocks of these handsome birds—sometimes hundreds strong—can be carried by an ascending thermal to a height at which, despite their size, they become scarcely visible from the ground.

Cranes are even more impressive to witness in their migrations than are the storks. Since remote antiquity, European cranes have been seen migrating from their breeding territories in Scandinavia, Lapland and northern Germany to Spain, North Africa, the Sudan and Ethiopia, where they spend the winter. Unlike storks, which are exclusively diurnal, cranes seldom make use of ascending air currents, and are much less dependent upon atmospheric conditions. They fly both during the day and at night, and cross the Mediterranean at its widest part—the eastern basin. Their measured wing-beat and flapping flight enables them to fly in the familiar and exciting V-shaped formation, whereas storks travel in formless flocks.

Crows, rooks (*Corvus frugilegus*) and jackdaws

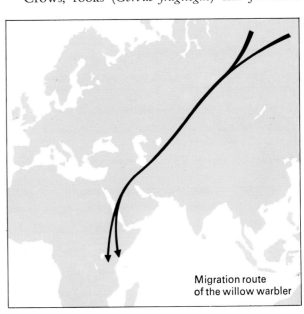

Migration route
of the willow warbler

(*C. monedula*) are often migratory in much of their range. During the winter, numbers of rooks, in particular, visit France, Germany, Belgium, the Netherlands and southern England from their breeding territories in eastern Europe. Gulls, ducks and starlings also undergo extensive migrations throughout Europe, and large numbers of smaller perching birds are also migratory. Warblers, thrushes, wheatears (*Oenanthe oenanthe*), finches and their allies often cover great distances despite their small size and delicate build. Their annual journeys may take them through Europe, over the Mediterranean and across the Sahara into tropical Africa, but, as they usually travel mainly at night and spend the day resting and feeding, their migrations are often overlooked, unless they are picked up as 'angels' on radar screens. In general, small and relatively secretive species—flycatchers, orioles, shrikes, warblers and other small, perching birds— tend to make their long migrations at night, while hawks, doves, swifts, swallows, crows, herons, storks, geese, ducks, gulls, and some seed-eating sparrows and finches usually travel by day.

Many birds of prey are migratory. Buzzards migrate from northern Europe to winter in Africa and southern Asia although, in Britain, they are sedentary and seldom wander over distances of more than 100 miles at most. Ospreys (*Pandion haliaetus*) are migratory throughout much of Europe

Facing page: the guillemot is a partial migrant from the Baltic, wintering on the European coast southwards to Spain; Above: osprey feeding its young. The osprey is a summer visitor to Western Europe; right: the wheatear, a small European migrant

as are Lesser Kestrels (*Falco naumanni*), Honey Buzzards (*Pernis apivorus*), Hobbies (*F. subbuteo*) and Red-footed Falcons (*F. vespertinus*)—which winter in South Africa. The migration routes of the osprey populations that winter in tropical Africa pass either over western Europe or across the Dardanelles and Asia Minor. These birds feed almost exclusively on large fish which, in winter, they would be unable to catch in the frozen lakes and rivers of northern and eastern Europe. The other predatory migrants are mainly insectivores. They tend to fly in large flocks—one flock of over a thousand Honey Buzzards has been described from Heligoland—and very many of them cross the Dardanelles every autumn.

Of all birds, the Falconiformes or accipiters, including the eagles, vultures and hawks, are among the most impressive and their migrations have attracted considerable interest among ornithologists. The falcons are powerful birds whose flight is exceptionally swift, the wings being pointed and the tail relatively short. Some species are solitary but others, such as the small Red-footed Falcon, are highly gregarious both on migration and when breeding. In the autumn they fly from Central Europe and southern Africa, where birds like the Lesser Kestrel are commonly to be found feeding on swarms of flying termites. Larger falcons, such

Left: the honey buzzard, a migrant that breeds in Europe and Northern Asia but winters in Africa, India and further east; below left: cream-coloured courser, a desert bird that moves toward the coast in summer; right: the gannet breeds on rocky islets around the coasts of Britain, Iceland and the Faeroes, from where many birds travel south as far as the coast of West Africa

as the Lanner Falcon (*Falco biarmicus*) feed mainly on small mammals and song-birds. In North Africa, song-birds are preyed on during their own migratory flights across the desert, an additional hazard of the trans-Saharan journey.

By no means all migrant birds undertake extensive journeys. Some desert birds, such as the Cream-coloured Courser (*Cursorius cursor*) and Bar-tailed Desert Lark (*Ammomanes cincturus*), which nest in the northern Sahara in early spring, move north in summer, because the region in which they breed is too hot and dry for them at that time of year. But they only go as far as the coast of Morocco and Tunisia. During the breeding season, of course, all birds are restricted to the territory surrounding their nests but, in winter, many of them tend to become nomadic and to disperse. The winter dispersion of Black-headed Gulls (*Larus ridibundus*), for instance, is very extensive, unlike that of Herring Gulls (*L. argentatus*). Other sea-birds that disperse in winter are guillemots (*Uria aalge*) and gannets (*Sula bassana*).

One apparently strange pattern of migration is the habit among many species of Eurasian birds of overwintering in the savanna regions south of the Sahara. This is the dry season there, during which the absence of rain, combined with grassfires, creates conditions that, superficially, appear conducive to exceptional barrenness. Nevertheless, towards the end of winter, when insects are scarce and the

vegetation is least productive, these birds manage rapidly to lay down stores of fat in preparation for their journey northwards in the spring. This paradox is probably more apparent than real. It is true that these migrants arrive each year in autumn at about the end of the rains, and thereafter witness a progressive desiccation of the herbage and some defoliation of the trees, but conditions are still good enough for them to accumulate the reserves that are necessary to enable them to cross the Sahara the following year. Clearly, the advantages of returning to temperate regions early in spring more than offsets the drawbacks of over-wintering during the dry season of the savanna. Moreover, when the rains do come, they are accompanied by a number of migrants, which are potential competitors, from the south. These include the Abdim's Storks, Carmine Bee-eaters (*Merops nubicus*), and nightjars, which inhabit the

northern savanna during the rains, after all the European visitors have departed.

Some tropical birds migrate regularly in relation to the occurrence of wet and dry seasons on either side of the Equator. Thus Abdim's Storks (*Sphenorhyncus abdimii*) breed in July and August, during the monsoon in the northern savanna regions of Africa, where they are regarded as a harbinger of rain. They move south in October and November, passing through the East African grass belts during the rainy season there, and inhabit the wet grasslands from Tanzania to the Transvaal at the time of the southern rainy season, which coincides with the dry season in the north. On the other hand, the Pennant-winged Nightjar (*Macrodipteryx vexillarius*), a very conspicuous bird in its breeding plumage, nests during the southern spring just before the rains and then migrates across the Equator as far north as

Below: grey heron, a partial migrant throughout its range. On the map are marked, from left to right, wintering areas, areas in which herons appear all year round and areas used for breeding only; bottom right: open-billed stork, a trans-equatorial migrant

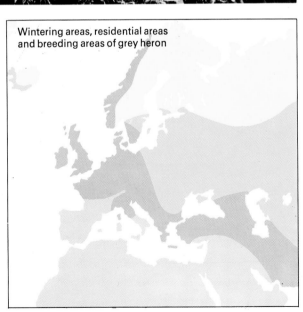

Wintering areas, residential areas and breeding areas of grey heron

Darfur in the western Sudan, where it moults.

It is especially important to the Pennant-winged Nightjar to follow the rains because it feeds mainly on winged termites which swarm when the ground is wet. Birds that migrate across the Equator thus get the benefit of two wet seasons per year. Other species which breed in southern Africa and regularly cross the Equator to spend their off-season in the north are the Rufous-cheeked Nightjar (*Caprimulgus rufigena*), the Amethyst Starling (*Cinnyricinclus leucogaster*) and the Open-billed Stork (*Anastamus lamelligerus*).

Another type of bird migration, more marked than mere local or trivial movements, is found in species that nest on mountains in summer, but fly down into the valley in winter. An example of this is the White-capped Water Redstart (*Chaimarrornis leucocephalus*) which lives along mountain torrents

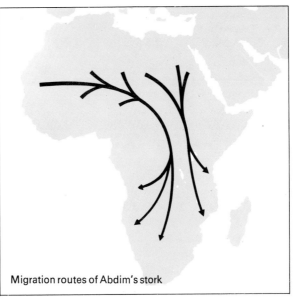

Migration routes of Abdim's stork

Above: Abdim's stork, an African species that breeds in the northern savanna at the time of the rains and later migrates across the equator, as shown in the map at left

41

and breeds in the Himalayas at 220–400 metres (7,000–13,000 feet), but winters at an altitude of 60–240 metres (2,000–8,000 feet). In general, migratory birds tend to breed in the coolest regions that they visit.

Less regular migrations

European Grey Herons (*Ardea cinerea*) provide an interesting example of migration patterns for they present almost every type of migratory movement within the same population. Some individuals are true migrants, radiating in winter throughout western Europe and moving particularly to France and Spain. Other birds are comparatively sedentary, while still others are vagrants that wander over considerable distances from the heronries in which they were hatched. If they did not do so, of course, nearby fishing grounds would soon become depleted.

The population density of most frugivorous (fruit-eating) and granivorous (grain-eating) birds depends upon the amount of food to be obtained in a given region. This is particularly true of species with a highly specialized diet based on the product of a single kind of plant. For instance, the Red Crossbill (*Loxia curvirostra*) depends entirely upon spruce for its nourishment. Its beak is wonderfully adapted for prying seeds from between the tight scales of the cones, the tips of both mandibles being twisted so that the tips overlap. Crossbills live in the upper branches of the trees, like little parrots, which they resemble somewhat in their gay plumage. They nest chiefly in the coniferous forests of Scandinavia and central Europe as well as in the Alps, the Massif Central of France, Spain and Greece. The density of crossbill populations is closely related to the spruce cone crops, and when the breeding season has ended, the birds begin to wander as soon as their

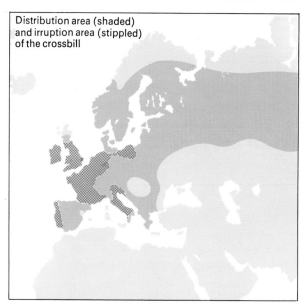

Distribution area (shaded) and irruption area (stippled) of the crossbill

Top left: Clark's nutcracker, a North and Central American species that feeds on pine seeds. Its only close relative, the Siberian nutcracker of Europe and Asia, top right, has a similar diet. Above left: the crossbill; left: its normal distribution (shaded) and irruption area (stippled)

42

food becomes scarce. Occasionally great numbers irrupt into western Europe, especially France and England. A crossbill invasion of Britain was chronicled by Matthew Paris in 1251. The movements of American crossbills are very similar to those of the European species. These birds are erratic nomads in winter, occasionally emigrating far beyond their normal territory, probably in response to food shortage.

The Siberian Nutcracker (*Nucifraga caryocatactes*) shows a similar behaviour pattern. Its food consists largely of Arolla pine seeds, which are much larger than the seeds of other coniferous trees. Within its beak, a cavity between the upper mandible and a projection of the lower is finely adapted for breaking the hard coverings of these seeds. Although Siberian Nutcrackers prepare for the winter by hiding a supply of seeds between roots in the soil or in rocky

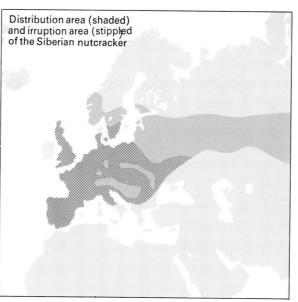

Distribution area (shaded) and irruption area (stippled) of the Siberian nutcracker

Left: distribution area (shaded) and irruption area (stippled) of the Siberian nutcracker. These birds are forced to emigrate in years when there is a shortage of pine seeds

Above: the capercaillie spends the summer in deciduous forest before migrating north to the taiga where it feeds on pine needles during winter; right: the snowy owl of the Arctic moves southwards to North America when lemmings and other rodents become scarce

crevices, their food supply is by no means constant from one year to another. A bumper year of cones is usually followed by several years when the yield is low or almost non-existent. To avoid starvation during very lean years, they leave the forests in which they usually spend the winter and fly westwards, sometimes in large flocks, across the plains of Russia and on to France and England. For centuries, the people of Poland, northern Germany and the Ukraine have considered nutcrackers to be portents of disaster. Some of the birds breed in their new homes the following spring, but few ever return to the original forests from which they emigrated. Like the nutcrackers, the Siberian squirrels are very dependent on Arolla seeds in the Siberian coniferous forest, and the number of squirrels sold each year at Siberian fairs closely reflects the Arolla cone crop.

In America, Clark's Nutcracker (*Nucifraga columbiana*), which feeds on the cones of White Fir and on Piñon, Sugar and Ponderosa Pine, also emigrates in times of food shortage. During the present century there have been four or five spectacular invasions from the mountain ranges to the coastal and desert regions nearby.

Much less is known about the movements of grouse, many of which are sedentary unless heavy snowfall prevents them from finding food. When this occurs, they migrate to regions where conditions are more favourable, chiefly to the great coniferous forests of the north. Thus the Willow Grouse (*Lagopus lagopus*) which breeds in the tundra, moves south in September to the edge of the forest. At this season, the birches and dwarf willows, on which it feeds during summer, are covered by snow so thick and hard that the birds are unable to break it. When the winter cold lessens, the Willow Grouse fly back to the tundra.

In addition to such regular migrations, however, population irruptions sometimes engender emigration in all directions—including north. At such times,

Willow Grouse have been seen flying in considerable numbers along the coast of the Arctic Ocean, and out over the frozen sea, where large numbers perish from starvation, thirst and cold. In autumn, the Capercaillie (*Tetrao urogallus*) leave the leafless hardwood forests where they can no longer find food, and fly north towards the coniferous *taiga* where they eat pine needles all the winter.

The irregular and sporadic movements of some birds may be related to shortage of food during a particular year. Many birds of prey rely upon fluctuating populations of lemmings for their food, and become widely dispersed when rodent numbers are low. Snowy Owls (*Nyctea nyctea*) migrate into

Canada and the northern United States on an average of once in four years when prey is scarce in their normal Arctic environment. A similar correlation is found between the numbers of lemmings and diurnal birds of prey, such as skuas (*Stercorarius* spp.), gyrfalcons (*Falco rusticolus*), and Northern Shrikes (*Lanius excubitor*), which also feed on them. Rough-legged Hawks (*Buteo lagopus*) often migrate through Germany when their rodent prey becomes scarce in Scandinavia, but these invasions are less regular in occurrence than those of the Snowy Owl. The irregular rainfall in desert regions forces birds, such as the Australian budgerigars, into a nomadic existence, as they search continually for areas where

rain has recently fallen and food is likely to be found.

The best-known example of an irruption is that of Pallas's Sandgrouse, mentioned earlier. This seed-eating species normally inhabits the steppes and deserts of western central Asia but, in the spring of certain years, the birds have migrated westwards in great numbers, attempting to breed in the new countries visited. There is some evidence of a return movement by the survivors in the following autumn. Large invasions took place in 1863, 1888 and 1900, and smaller ones in other years. In 1863 many of these birds reached Great Britain, Ireland, the Shetlands and Faroes, but although they succeeded in breeding in many places, the species never became established.

No doubt these emigrations were occasioned by food shortage, for the nomadic tribesmen have a saying that 'when the sandgrouse fly by, wives will be cheap'.

Similarly the Rosy Pastor or Rose-coloured Starling (*Sturnus roseus*) makes occasional visits to western Europe. Nesting over a vast area from south-eastern Europe to Turkestan and the Altai Mountains, it feeds chiefly on locusts and grasshoppers. When the locust swarms migrate, the Rosy Pastor leaves its breeding ground and disappears in pursuit of them. It winters as far away as Great Britain, France and North America, wandering nomadically for considerable distances at intervals which, however, can be fitted into more or less regular cycles. Large numbers lived in Hungary as recently as 1908, 1925, and 1932. Such irruptive phenomena are related to the irregular rainfall and movements of locusts in arid regions, for the Wattled Starling (*Creatophora cinerea*) of southern Africa responds in much the same way, following and breeding alongside locust swarms. Flocks of wintering White Storks and other birds also follow locust swarms in Africa. Similar nomadic habits are found among the water-birds of the Australian desert such as the Banded Stilt (*Cladorhynchus leucocephalus*) and Black-tailed Water Hen (*Tribonyx ventralis*), which fly from the site of one rainstorm to that of another, the Purple-crowned Lorikeet (*Glossopsitta porphyrocephala*), which seeks out those regions in which eucalyptus is in flower, and the Masked Swallow Shrike or Wood Swallow (*Lanius nubicus*). The last named is truly migratory as well as being nomadic throughout part of the year.

From all this, it might seem that irruptions are always caused directly by food shortage, but this is not so. The Bohemian Waxwing (*Bombycilla garrulus*) which breeds in the forests of northern Europe and Asia, engages in extensive migrations to central Europe each winter. Only part of the population leaves its summer territory, however, and at intervals of approximately ten years, great irruptions of

Below: white storks resting on their southerly migration into Africa. Many storks return annually to the same nests, invariably built on used chimneys and rooftops

waxwings occur, which are quite different from the normal migratory movements. Beginning much earlier in the autumn, they extend over considerably greater areas than those covered by the partial migrations or nomadic wanderings of the intervening years. Few of the migrants return and, although some of them remain in the areas they have invaded, they never succeed in establishing themselves. Some ornithologists have tried to attribute these irruptions, without success, to fluctuations in the annual crops of rowan berries, the waxwings' favourite diet in winter, but there seems to be no correlation between invasions of waxwings and an abundance or scarcity of these berries, although food does play a rôle in the erratic migrations between invasions. When there is an abundant crop of berries in the north, waxwings spend the whole winter there and do not migrate to central Europe.

The ten-year cycle of irruptions by waxwings is not understood. It appears to be related to some internal rhythm controlling increases or decreases

The Bohemian waxwing breeds in the forests of northern Europe and Asia but migrates to Central Europe in winter. Large irruptions of the waxwing, the reason for which is not yet established, occur approximately every ten years

in population numbers and not to be influenced directly by environmental factors. Certain mammals pose similar problems for which there are, as yet, no valid explanations. On the whole, irruptions leading to large-scale emigrations are clearly distinguished from true migration, although so many intermediate stages are to be found that the two are best regarded as extreme cases of one continuous spectrum of migratory behaviour.

North America

So far, this chapter has centred mainly on the migration of Eurasian birds: North American species, however, have to contend with similar nutritional problems in winter. In the New World there are no barriers to migration, such as those imposed by the Mediterranean and the Sahara. It is not surprising, therefore, to find that the North

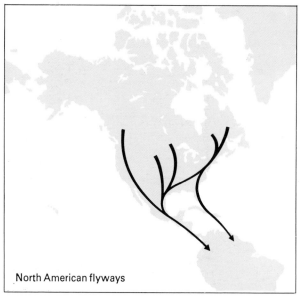

North American flyways

Above: Canada geese, or 'honkers', migrating through the United States where they feature as one of the principal quarries of American 'duck-hunters'; left: the North American flyways

Right: the rose-breasted grosbeak, a North American species, winters in Central and South America; below: white-fronted geese are a familiar sight in Britain during winter after nesting in the Arctic tundra

American bird fauna includes a large proportion of migrants. In addition to swallows, martins, hawks, crows, geese, ducks and perching birds, North America is visited each year by several species of birds with tropical affinities, such as tanagers and hummingbirds, whose counterparts in the Old World are extremely few.

North American waders and other shore birds, like their counterparts in Europe and Asia, are great migrants and winter in the warmer zones of the eastern hemisphere. The American Golden Plover (*Pluvialis dominica*) travels further than most, migrating from its nesting territories in the tundra of Alaska and northern Canada to winter in the pampas of Argentina. Many of the pigeons are migratory, and the extinct Passenger Pigeon (*Ectopistes migratorius*), whose flocks once darkened the sky, used to nest in the deciduous forests of the United States and spend the winter in a relatively small area in the southern Ohio valley. The Rose-breasted Grosbeak (*Pheucticus ludovicianus*) breeds in the north-eastern United States and southern Canada, wintering in Central America and the north-western corner of South America, while the American Golden Plover has a breeding range extending north to Alaska and migrates to the Argentine in winter. The bobolink or Rice-bird, a native of the same general region as the Rose-breasted Grosbeak, crosses the Equator to the southern tropics in Paraguay and Brazil.

The migration of North American birds to winter quarters in the south, follows a number of different routes, or 'flyways', although the general direction is necessarily south or south-eastwards. On the western side of the continent, the most direct route is along the coast and overland by way of Central America. This is known as the Pacific Flyway: it is used by many perching birds, geese and shore birds. The Central or Great Plains–Rocky Mountains Flyway starts in Alaska and follows the eastern side of the Rocky Mountains down to Mexico and Central America. A large number of duck species use this route, including wigeon (*Anas penelope*), Common Pintails (*A. acuta*) and redheads (*Aythya americana*).

The Mississippi Flyway, the most important of all, runs from Alaska and the Mackenzie River to the Great Lakes, and then follows the valley of the Mississippi to the Gulf of Mexico. The main route is fed by a number of lateral branches, from east and west, including the valleys of the Ohio and Missouri Rivers. This path is taken by many waterfowl, especially Canada Geese (*Branta canadensis*), mallards, and Common Pintails, large numbers of perching birds, including the American Robin (*Turdus migratorius*) and Myrtle Warbler (*Dendroica coronata*), swallows, catbirds (*Dumetella carolinensis*), and flycatchers. Lastly, the Atlantic Coast Flyway runs from Greenland and north-eastern Canada to

Above: the American golden plover migrates from nesting territories in the tundra of Alaska and nothern Canada to winter in the Argentine pampas; right: migration routes of American and Eurasian populations

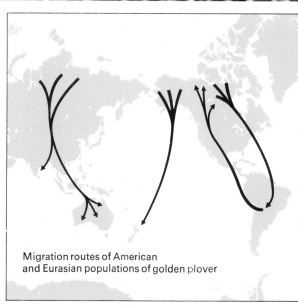

Migration routes of American and Eurasian populations of golden plover

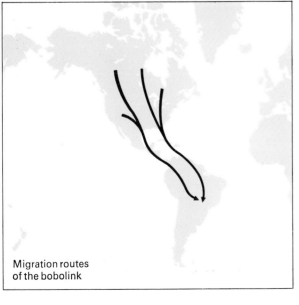

Migration routes
of the bobolink

*Left: the migration
route of the bobolink
follows the American
flyways; right:
migration route of the
blue goose*

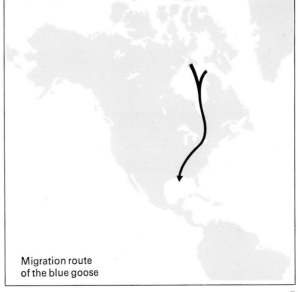

Migration route
of the blue goose

Delaware and Chesapeake Bays where immense numbers of wildfowl stop for the winter. Many birds of prey and perching birds, however, continue from Florida to Cuba, the West Indies and on to South America. This path is also followed by the bobolink, various cuckoos, warblers and thrushes, but it is used less than the other flyways, no doubt because it crosses so much sea. On the other hand, the Golden Plover and a number of water-birds fly direct from Nova Scotia and Labrador over the Atlantic Ocean to the coast of Brazil; but no land-birds follow this dangerous route.

South America

South America provides winter quarters for northern birds to a lesser extent than Africa does. It has no northern sub-tropical zone like that which accommodates so many migrants in northern Africa, and it seems probable that the equatorial rain-forest belt offers a greater barrier than in Africa to trans-equatorial migrants. Many birds that breed in the temperate regions of southern South America move northwards towards the Equator for the winter, as happens to a lesser extent in southern Africa, and there are several trans-equatorial migrants in South America that follow the rains, like their counterparts in Africa, the Abdim's Stork and other African species.

Australasia

Australia is the wintering ground of a large number of migrants from the northern hemisphere, mainly plovers and their allies which nest in northern Asia and make enormous communal flights along the coast and down the Malayan peninsula, as well as Fork-tailed Palm Swifts (*Tachornis squamata*) and Spine-tailed Swifts which make regular flights in large flocks from China and Japan. In addition, before the Australian winter, large numbers of sea-birds, including Wilson's Storm-Petrel and the

Southern Skua (*Stercorarius maccormicki*), fly from the south to avoid the dark and bitter winters of the Antarctic continent. Some of these are also received by New Zealand, although that country lies beyond the world's great bird migration routes and none of her resident perching birds is migratory.

Among the many Australian birds that perform local or, occasionally, altitudinal migrations are several species of honey-eaters which have to search for the blossoms from which they draw their nourishment. Thus, the Scarlet Honey-eater (*Myzomela sanguinolenta*) remains in New South Wales if the bottle-brush trees flower abundantly, but leaves the region if they do not, while the movements of the Painted Honey-eater (*Conopophila picta*) between Victoria and the Northern Territory are controlled by the fruiting of the mistletoe on which it feeds. The Australian Budgerigar (*Melopsittacus undulatus*), Black Cockatoo (*Calyptorhynchus funereus*), Grey Fantail (*Rhipidura fulginosa*) and Golden Whistler (*Pachycephala pectoralis*) are among the birds which move to moist, hilly regions during the summer season when the lowlands are extremely dry. Bird migrations thus range between the daily nomadic flights of sandgrouse between their gritting grounds and water to the dramatic 22,500-kilometre (14,000-mile) oceanic journey of an Arctic Tern, banded on the White Sea coast about 200 kilometres (125 miles) from Murmansk and captured alive the following year by a fisherman, a few kilometres south of Fremantle, in Western Australia. They include the short seasonal migrations of the Black-capped Chickadee (*Parus atricapillus*), Clark's Nutcracker, and Rosy Finch (*Leucosticte arctoa*) of North America—which merely descend from exposed mountain ridges to the sheltered valleys below—to the great spasmodic irruptions of crossbills and Siberian Nutcrackers.

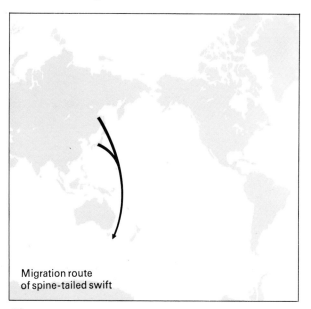

Migration route
of spine-tailed swift

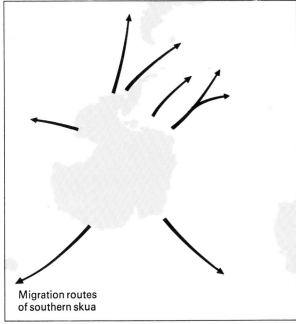

Migration routes
of southern skua

*Above, left and left:
the wings of the
wandering albatross are
constructed to facilitate
dynamic soaring in
strong winds; above,
centre: migration
routes of the southern
skua; above: southern
skua in its dark phase.
This species flies north
to avoid the Antarctic
winter; far left:
migration route of the
spine-tailed swift, which
flies in large flocks from
the Far East*

Within this vast range, abnormalities do sometimes occur. Some hardy species, especially of finches, that normally perform relatively short journeys, have been known to perform migrations in the reverse direction to that normally taken. Such occurrences may be due to unusual weather conditions. Ducks, too, which have wintered in their breeding territory, occasionally move northwards in spring, probably attached to flocks of passing migrants.

Day and night, summer and winter, year in and year out, birds are migrating in some or other part of the world. A few species may undergo non-stop flights of great length but, in general, migration is a fairly leisurely process that, nevertheless, proceeds relentlessly until at last the journey has ended and reproduction can take place. It is not the extent of some of these journeys, so much as the magnitude of the phenomenon, that is finally so impressive.

Bats

Another important group of flying animals is the bats, whose life histories, since they are mammals, differ significantly from those of birds. In birds, courtship, egg-laying and the feeding of the young follow in quick succession. Among mammals, on the other hand gestation takes so long that courtship normally occurs at a different season from that in which the young are born. This raises a difficulty for the bats of temperate regions, whose gestation period lasts for about two months. If they were to mate some weeks after awakening from hibernation, much of the summer would be over before the young were born. On the other hand, bats could hardly be expected to mate in the depths of winter, when they are hibernating. The problem is overcome as follows: sperm production and copulation take place in the autumn before the animals hibernate, but the females store the living spermatozoa until the early spring, when fertilization takes place. In some species, spermatozoa are also stored by the males for months after they have been produced.

In the northern countries of the world, the insects upon which bats depend for food die off at the approach of winter, or hide away so that the bats of those regions must either hibernate or else move to warmer latitudes as swifts and swallows do. The species that inhabit caves, hollow trees, belfries and similar dark shelters, for the most part, hibernate in them, or where the cold is severe, seek out caverns sufficiently deep to maintain at all times a temperature above freezing point. In the eastern regions of the United States, such retreats are relatively few and far between, except in limestone districts where there are often extensive subterranean cave systems. The many individuals that aggregate in these during the winter, therefore, migrate from surrounding districts to which they return again in the following spring. Such migrations are comparatively short.

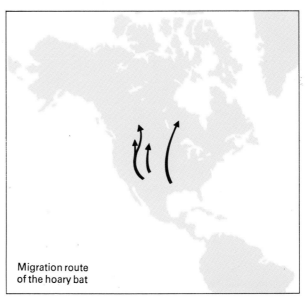

Migration route
of the hoary bat

Left: migration route of the hoary bat; right: a European long-eared bat eating an insect. This species migrates up to 260 km; below: Mexican free-tailed bats leaving the Carlsbad Caverns, Texas, at dusk. This species has been known to migrate more than 12,800 km

The species of bats that roost in trees, among the leaves or against the trunks, on the other hand, do not normally enter caves, but undergo extensive migrations to the south, which are comparable with the seasonal migrations of insectivorous birds.

The Red Bat (*Lasiurus borealis*) and Hoary Bats (*L. cinereus*) of North America, for example, have long, narrow wings that endow them with such powerful flight that they are able to migrate to the southern parts of the United States for the winter. Both species are also occasional autumn visitors on Bermuda, some 1,000 kilometres (650 miles) south-east of New York, and a Red Bat has been captured 385 kilometres (240 miles) off Cape Cod in August. Undoubtedly large numbers must migrate across the sea, although they probably follow the coast as far as possible. The Silver-haired Bat (*Lasionycteris noctivagans*) of North America also performs more or less regular annual migrations while the Mexican

Greater horseshoe bats hibernating in a cave. As with many species of birds, bats often migrate in flocks of the same sex

Free-tailed Bat (*Tardarida mexicana*)—so abundant in the Carlsbad Cavern—has been shown to migrate over 12,800 kilometres (8,000 miles) to central Mexico.

Compared with small birds and with other small mammals, bats are long-lived creatures and may migrate several times during their lives. The distance between summer and winter quarters varies considerably according to the species and individuals concerned. In the case of the European Long-eared Bat (*Plecotus auritus*), the maximum is about 260 kilometres (160 miles) and the minimum is less than a kilometre, with an average between 40 and 80 kilometres (25 and 50 miles). The distances travelled by noctules (*Nyctalus* spp.) are usually greater than this, while a Pipistrelle Banded at the end of June 1939 in the province of Dnepropetrovsk, USSR, was recovered nine weeks later, near Plovdiv in Bulgaria, at a distance of 1,150 kilometres (720 miles). In 1935, of 600 noctules banded during hibernation in a cave near Dresden, three were recovered—one near Hanover, a second from Susk in Poland and the third from Lithuania, a distance of nearly 750 kilometres (470 miles). In these animals, both migration and hibernation therefore take place.

Long-fingered Bats (*Miniopterus schreibersii*) and Greater Horseshoe Bats (*Rhinolophus ferrum-equinum*) often hibernate in caverns far from their usual summer areas and the same is true of the Lesser Horseshoe Bats (*R. hipposideros*) and White-edged Bats (*Pipistrellus kuhlii*). Thousands of Frosted Bats (*Vespertilio murinus*) spend the summer in the old ramparts of Aigues-Mortes, near the mouth of the Rhône but abandon them in the autumn. Presumably these caverns are not suitable for hibernation and are abandoned by the end of the year for less open retreats where, perhaps, the temperature remains more constant. But there is no evidence that European bats winter in North Africa as many birds do. Bats have, however, occasionally been observed in daylight migrating in Europe in company with swallows and other birds.

The autumnal migration of Eurasian and North American bats is carried out in a leisurely manner from mid-August until November, but is at its height in September. No doubt some of the bats of southern South America migrate northwards at the approach of the winter there, but little is known of the movements of bats in that part of the world.

The fruit-bats or flying foxes of the tropics are known to make fairly regular mass migrations in search of ripening fruit. Most of these bats are social in habit and assemble by day, often in vast numbers in favourite roosting places. One such locality is at Kampala, Uganda, where hundreds of thousands of bats festoon the branches of the trees during the daytime. Fruit-bats migrate to South Africa during the southern summer and fly northwards down the Nile at the season of the rains. In Australia, flying-foxes migrate periodically from Queensland to New South Wales. At times, the Grey-headed Fruit-bats (*Pteropus poliocephalus*), Gouldian Bats (*P. gouldi*) and Spectacled Bats (*P. conspicillatus*), in particular, do considerable damage to fruit-trees in northern and eastern Australia. The smaller Red Flying Fox (*P. scapulatus*), which feeds mainly on flowers, is less regular in its migrations, for these are conditioned largely by the somewhat uncertain flowering of eucalyptus and other trees, on the blossoms of which they feed. In contrast, the movements of the Grey-headed Fruit-bat are determined more by the seasonal ripening of small fruits, especially wild figs, which comprise the bulk of its food. Nevertheless, the two species are often found together, roosting in the same trees during the day.

As in many birds, migrating flocks of bats sometimes consist entirely of animals of the same sex. This segregation indicates, without doubt, that migration is concerned with feeding rather than with reproduction. By following the twilight northwards, as the days lengthen in spring and summer, insectivorous bats are able to feed upon the myriads of nutritious insects that have recently emerged from their winter quiescence. Their seasonal migrations, like those of birds, provide nutritional advantages both for the parents and for their young, and also enable them to escape the cold and dark of the winter.

Arthropods

The earlier discussion of animal migration suggested that trivial and nomadic movements, when directed and refined by the acid test of natural selection, may evolve into true, to-and-fro, migration. Such migration can occur one or more times during the lifetime of an individual, as in the case of swallows and cuckoos. Within a population of short-lived animals, like locusts and butterflies, on the other hand, the complete cycle may not be completed until after the passage of three or four generations. Animals whose populations undergo periodic fluctuations in numbers tend to emigrate at the peak of such irruptions, but may be quite sedentary in their habits at other times. Their emigration is regarded as being the first stage of a true migratory movement even if, in most cases, there may be few, if any, survivors to complete the return journey.

Despite their mobility, trivial movements and random dispersal are common among flying animals, including birds and bats. So many small insects, both winged and wingless, not to mention spiders and mites, are carried by chance air currents high into the upper atmosphere that it is no exaggeration to speak of 'aerial plankton' by analogy with the dense, floating, pelagic life of seas and lakes. Although their dispersal is random, in that these animals are at the

Newly hatched spiderlings in preparation for 'ballooning'

A 'money' spider greatly magnified. Even adults of such small species become dispersed by 'ballooning' and may be transported considerable distances attached to threads of web

the thread snaps near its hold.

Often during the late summer and autumn months, on quiet and hazy days after cold nights, the air is filled with shining strands and threads of gossamer—the silk produced by the migrating spiders that have attempted to fly, and failed. The prosaic translate 'gossamer' as 'goose summer' in reference to the fanciful resemblance of the fragile skeins of silk to the down of geese which the thrifty housewife causes to fly when she renovates her feather beds and pillows. However, gossamer translated as 'God's summer' refers to the legend that this gossamer is the remnant of Our Lady's winding sheet which fell away in these lightest of fragments as she was assumed into heaven.

Sometimes one sees a field or meadow carpeted with silk, and a host of little spiderlings spreading their lines in vain attempts to fly. On the other hand, many are successful. In 1839, Charles Darwin

mercy of the winds, the distinction between active and passive migration is apparent rather than real. It is now known that strongly flying locusts are virtually as dependent on the wind as are feeble fliers like aphids and thrips.

Much of the aerial plankton is made up of individual animals that, engaged in trivial locomotion, have been swept up and carried away by a sudden gust of wind. Such individuals, dispersed passively, are sometimes referred to as 'vagrants'. Their transport may have the same ecological effects as migratory movements but, behaviourally, it is quite distinct.

Among the many remarkable traits of spiders, none has excited greater interest, nor produced more fantastic speculation, than that of 'ballooning', the phenomenon in which small spiderlings are carried aloft, attached to threads of silk. Much of the adventure and risk in the life of the spider is crowded into the first few days of freedom when the young spiderlings, having first left their egg sac, wander over the stems of plants and up the leaves of grasses, stringing their threads as they go. Soon a tangle of webs springs up, crossing in all directions and covering the vegetation. When the young spider has reached the summit of the nearest promontory—a weed, a bush, or a fence—it turns to face the wind, extends its legs so that it appears to be standing on tiptoe and lets air currents carry the silk from its spinnerets. When the friction of the currents against the threads exerts sufficient pull, the spider loosens its hold and usually sails away: at the take-off at least, it is dragged backwards. Sometimes, after take-off, the spider climbs rapidly to the middle of its thread, which then sweeps forward and becomes doubled. Less frequently, the spider makes a forward start. This method is employed by smaller spiders which make a weak attachment to a support, and allow themselves to be blown outward and upward until

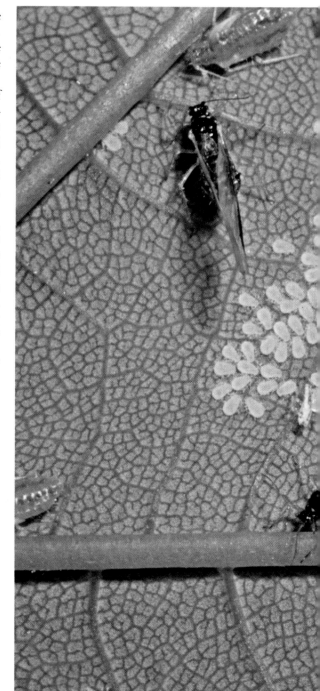

recorded the arrival on HMS *Beagle* of 'vast numbers of a small spider, about one-tenth of an inch in length, and of a dusky red colour' when the ship was about 95 kilometres (60 miles) from the coast of South America. Ballooning is, without doubt, an important factor in the distribution of many species all over the world, for tiny astronauts have been known to alight on the rigging of ships hundreds of kilometres from the nearest land. Nor is it confined to any particular season. In Britain, aeronautic dispersal of immature spiders takes place mainly in summer, and of adult 'money' spiders chiefly during the colder months, although temperature is the most important micro-climatic factor and their ballooning is inhibited during unfavourable weather.

That spiders undertake such airborne migrations has been known since the time of the ancient Greeks, for Aristotle believed that they could 'shoot out' their threads, and Pliny recorded that, 'In the year that L. Paulus and C. Marcellus were consuls, it rained wool.' Chaucer, also, mentions the possible influence of weather on gossamer:

> Sore wondress some on cause of thonder
> On ebb and floud, on gossamer, and on mist.

As late as 1664, Robert Hooke, the celebrated scientist, suggested to the Royal Society that, 'it is not unlikely that those great white clouds which appear all the summertime may be of the same substance' as gossamer. This may explain the derivation of the nursery rhyme in which the old woman is tossed in a basket to sweep cobwebs from the sky.

Although ballooning is entirely passive, spiders having no control over the direction nor, probably, over the duration of their flights, large numbers are undoubtedly carried aloft to considerable heights.

Winged and wingless female aphids on a sycamore leaf. Aphids are important members of the aerial plankton

Phoretic mites on the underside of a dor beetle which provides them with free transport

They have been recorded at 4,570 metres (15,000 feet) over Louisiana. Aerial movement by spiders is therefore migratory, since it carries the animals away from the territory or habitat they formerly occupied. Ballooning has made possible the distribution of spiders all over the world, from oceanic islands to the bleak cliffs of Mount Everest at an elevation of 6,700 metres (22,000 feet).

Members of most families of spiders, such as jumping spiders, wolf-spider, crab-spiders and orb-web spiders become astronauts only when they are immature, but the tiny black 'money' spiders embark on ballooning flights when they are adult also. Neither the numbers engaging in aerial dispersal, nor the phenomenon itself, is related to the numbers of species present in a particular region, or to their densities on the ground. Furthermore, only certain species ever become dispersed in this

way and, among these aeronauts, the peak of aerial dispersal often fails to coincide with the population density. It is obvious, therefore, that ballooning must be a normal event at a definite phase of the life-cycle of the aeronautic species. When the individuals of these species are in the ballooning phase, migratory behaviour is stimulated by suitable weather conditions.

Migration tends to be especially characteristic of spiders that occupy temporary and varied habitats. In contrast to permanent habitats, such as rivers, lakes, heaths, marshes and perennial plants, temporary habitats include dung, carrion, fungi, plant debris, temporary ponds and annual plants. The same is true of mites, of which members of certain families, in their nymphal stages, are regularly dispersed with the aerial plankton. Like springtails, they have a tendency to make vertical leaps when they feel a puff of wind, and this increases their chances of being carried upwards.

Animals of the aerial plankton are of three kinds, of which spiders are one. In proportion to their surface area, even tiny spiders are too heavy to float without the aid of support, and are only able to do so by producing silk threads which carry them aloft. Much smaller animals, however, such as some mites and springtails, whose weight is insignificant in proportion to their surface area, are able to keep floating like particles of dust, without any additional support. Many of them are normally found in damp environments, and it is not really understood why they should not become desiccated in the atmosphere. Perhaps because they drift with the wind and not through the air, they may become surrounded by a layer of moisture that is not removed, and so experience little tendency to dry up. In addition, such animals are most easily transported by storms and hurricanes when the upper atmosphere is very humid.

The largest group of aerial plankton includes small winged insects such as aphids, whitefly, thrips, fruit-flies, frit-flies, mosquitoes and midges, which are regularly dispersed by wind. In some cases, their chances of being carried into the upper air are increased by specific behaviour patterns. For instance, aphids and frit-flies will climb to the highest portion of the vegetation on which they are feeding, face the sun and then take off, flying upwards into the air. This reaction is especially marked in younger insects, and appears to be dependent upon favourable weather conditions.

Some arthropods, wingless yet too heavy to be carried into the air without other support, are 'phoretic'. That is, they attach themselves to another animal which provides them with free transport until they let go and drop off. Since they are completely harmless to their 'hosts', they are in no way parasitic. For example, the nymphs of uropodid mites attach themselves to passing insects with the aid of a thread of secretion from the anal opening. This dries, and hardens to form a kind of cement. The attached mite is thus free to move about and can escape, at will, by tearing itself loose from the attachment. Acarideid mites, on the other hand, produce a hairy deutonymph armed with discs by which it is able to attach itself to the legs of insects and other arthropods such as long-legged, mobile harvestmen. The curious little false-scorpions, commonly to be found in leaf litter, under the bark of trees and in similar localities, attach themselves to the legs of flies and other arthropods by means of their enlarged claws. The phenomenon is restricted to mature females. Phoresy is rarely found in male false-scorpions and never in the juvenile stages. It is significant that the phoretic animals are physiologically adapted for their journeys by acquiring greater resistance to lack of food and to desiccation,

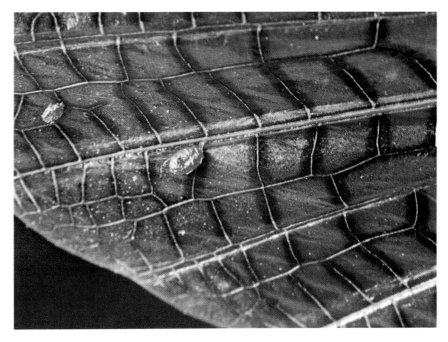

Above: phoretic mites attached to the wing of a locust; below: swallowtail butterflies on migration in Ethiopia pause to drink from a puddle

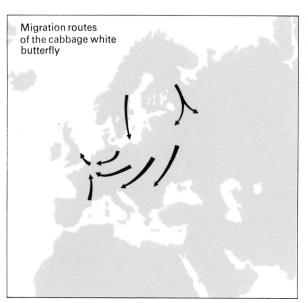

Migration routes
of the cabbage white
butterfly

Left: migration route of the cabbage white butterfly; right: small tortoiseshell butterfly, a well-known migrant in Europe; far right: migration routes of the monarch butterfly

than are the other developmental stages of mites and false-scorpions.

Phoresy is always a migratory movement and, with false-scorpions, it would seem to be the main, if not the only, regular form of migration. The species that are commonly phoretic tend to be those that inhabit vegetable débris, manure heaps and dry seaweed cast upon the shore, while soil dwellers and those that live beneath the bark of trees are not phoretic. That the phoretic habit has been evolved as a means of colonizing new habitats is shown by the fact that it is found mainly in the mature, presumably fertilized, females of species that normally exploit temporary habitats.

Far more than by observations of chance emigration dictated by wind and weather, the problem of migration has been drawn to the attention of entomologists by the appearance of large numbers of insects, flying for hours or days in one definite direction. At first, such movements were thought to be quite local affairs covering, at most, a few kilometres—as, indeed some of them really are—but the number of records increased so rapidly, and the distances traversed were sometimes found to be so large, that it came to be realized that insect migration is an extensive and important phenomenon. Although the winged adults of some insect species survive but a few days, many live for several weeks or even months. It is among these long-lived insects, including butterflies, dragonflies and locusts, that migration is most conspicuous.

Sometimes swarms of butterflies in Africa are so vast that they stretch for perhaps 25 kilometres (15 miles) with stragglers extending several kilometres more in each direction. Two species of white butterflies are outstanding migrants south of the Sahara. One of these ranges throughout the continent except in the western tropical forests, the other is common in eastern Africa. The larvae of both feed on the leaves of spurge, and their food plants

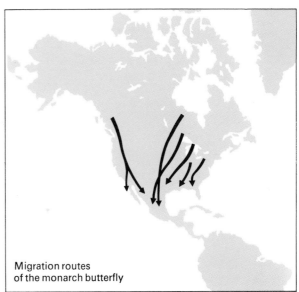

Migration routes
of the monarch butterfly

are sometimes completely defoliated by thousands of caterpillars. Another species is a regular migrant in West Africa, where it moves southward early in the rains—that is, in February to May—returning northwards from October to December. Several other species have been recorded migrating in different parts of the Continent.

More is known of the migrations of butterflies and moths than of any other insects, since Lepidoptera are conspicuous and nearly always move within six metres (20 feet) of the ground. Most migratory species in tropical countries are found in arid and semi-arid regions, where rainfall is sporadic. Under such conditions, insects are especially prone to migratory movements when conditions become locally unfavourable.

The migration of butterflies, like that of birds but unlike locust migration, is not to any great extent determined by the wind. The migrants usually keep direction by steering by the sun for kilometre after kilometre. Most people residing in Britain do not experience the sight of large numbers of migrating butterflies more than once or twice in a lifetime although some common species of the British Isles, such as the Red Admiral (*Vanessa atlanta*), Small Tortoiseshell (*Vanessa urticae*), and the Clouded Yellows (*Colias croceus* and *C. hyale*), regularly move northwards from the continent of Europe during the spring and fly southwards again during the autumn.

In warmer regions of the world, mass migrations of Lepidoptera are comparatively frequent. Two North American butterflies, the Monarch (*Danaus plexippus*) and the Painted Lady (*Vanessa cardui*), have been studied extensively. Although they seldom or never breed in Britain, these are among the 18 butterfly species and nine species of hawk-moths that regularly migrate to this country. The Monarch, or Milkweed Butterfly, is a large chestnut-brown insect with dark veins on the wings. Its home is in

63

North America, but it is a remarkable migrant and occasionally spreads both west and eastward. Over 200 individuals have been captured or seen in Britain during the past century although its food plant, milkweed, does not occur naturally in Europe. In America, the Monarch's distribution in summer extends over the United States and southern Canada, occasionally reaching as far north as the Hudson Bay. As autumn sets in, however, the butterflies in the north begin to migrate steadily and slowly southwards, joining up in the small bands that occasionally increase to form gigantic swarms. In 1885, a swarm settled in New Jersey that covered every available leaf and twig over an area of about 320 kilometres (200 miles) wide and over five kilometres (2 miles) in depth. The following morning they had all disappeared.

Monarchs spend the winter in semi-hibernation roosting on trees in Florida, California and Mexico. They are mostly inactive at this time of year but, on warm days, occasionally fly around in the sunshine. Thousands of butterflies may occupy the same tree, which, for unknown reasons, is selected by the migrants, year after year. Winter quarters are abandoned as spring approaches, and the Monarchs begin the long return flight to the regions from which they emigrated the previous year. Here they lay their eggs and die. Two or three generations are then passed before the following autumn, when the annual southward migration begins again. Those

intermediate generations do not migrate however, the entire two-way journey over 5,600 kilometres (3,500 miles) being accomplished by the same enterprising individuals.

Proof that an insect has migrated over great distances is difficult to obtain except in the case of the Monarchs, many of which have been marked and recovered. In 1960, however, a small pyralid moth captured in England was found to contain radio-active particles of a type derived from the atomic bomb exploded in the Sahara a month earlier. This suggested very strongly that this small insect had migrated for some 2,400 kilometres (1,500 miles).

Most of the Monarchs that migrate to Britain, do so in September. They are probably representatives of the southward migration in America that, borne on the prevailing westerly winds, have been accidentally blown across the Atlantic. Over 100 years ago, the Monarch spread westwards across the Pacific to Hawaii, New Zealand, Australia, Borneo and Hong Kong where it became established in several places.

Another famous migrant from North America is the Camberwell Beauty (*Nymphalis antiopa*), first recognized in England in 1748. The Camberwell Beauty is one of Britain's rarest butterfly visitors. The first two to be collected were taken about the middle of August, 1748, in Cool Arbour Lane near Camberwell and were recorded in *The Aurelian* of Moses Harris, a folio volume of most beautiful

Below: convolvulus hawk-moth, an immigrant to Britain each year; facing page: part of a huge migratory swarm of monarch butterflies resting on a tree trunk

execution, full of accurate and valuable information. Published in 1766, *The Aurelian* is a rare work which, for generations, served as a guide for students and collectors.

The Bath White (*Pontia daplidice*) has been known to occur in Britain for over 250 years, but it is always as a great rarity. An irregular vagrant in the United Kingdom, it migrates regularly through the passes in the Pyrenees between France and Spain. Mass irruptions have been recorded in the High Atlas Mountains of Morocco during the spring, and in Eritrea during October. The Painted Lady breeds in Britain every summer, but seldom, if ever, survives the winter. Immigrants normally arrive at the end of May and in early June, but the numbers vary greatly from year to year. This species, one of the most widely distributed of butterflies, has been found in every continent except South America.

In winter, the Painted Lady breeds along the northern fringe of the Sahara. Then it moves northwards as far as northern England and Germany, reaching Iceland and northern Finland in years of abundance there is evidence of a return flight to the south in autumn. Painted Ladies have been seen in thousands flying through Egypt and across the Mediterranean.

There are many more species of moths than of butterflies and some of the larger are well-known migrants. Half of the 18 British hawk-moth species arrive annually as immigrants, and do not survive the winter. They vary in abundance from the Hummingbird and Convolvulus Hawk-moths (*Agrius convolvuli*), which come over in fair numbers each year, to the rare Spurge Hawk-moth (*Hyles euphorbiae*), of which less than 20 individuals have been recorded during the past 100 years. The Hummingbird Hawk-moth (*Macroglossum stellaturum*) is found

throughout southern Europe, North Africa and Asia as far as Korea, Japan, and southern India. Many thousands regularly migrate northwards in spring, returning to the south in the autumn months.

The Urania Moth (*Urania leilus*) occasionally undergoes widespread migrations throughout tropical America, the causes of which are, as yet, quite unknown. In Australia, the Bugong Moth (*Euxoa infusa*) aestivates during the hot season in the hills, and moves back to the plains at the end of summer to breed there during the colder months. In the past, the Aborigines used these insects as a source of food. They knew all the clefts and caves in which the moths tended to aggregate, and collected them in large numbers after stupefying them with smoke.

Apart from Lepidoptera, the more important migratory insects are locusts and other grasshoppers. The young and larval stages of most insects are dedicated to feeding and growth so it is natural that dispersal and migration should be undertaken mainly by the winged adults. In the case of grasshoppers, however, population irruptions may engender migration in the hopper instars as well. The phase of restless activity that culminates in migration usually appears among older nymphs and in the young adults soon after their final moult. We have already seen that one of several basic strategies are available to living organisms by which they can survive unfavourable climatic conditions. They can

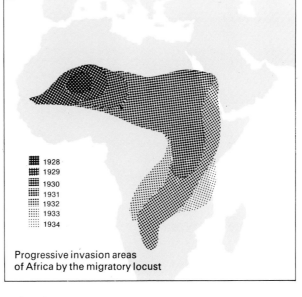

1928
1929
1930
1931
1932
1933
1934

Progressive invasion areas
of Africa by the migratory locust

Over a period of thirteen years swarms of migratory locusts invaded Africa, followed later by the red locust

either hibernate in winter in a state of suspended animation, aestivate likewise in summer, carry on as usual, or migrate. The suggestion has been made that migration itself may be an example of arrested development, like hibernation or aestivation, since migrating insects—be they aphids, butterflies or locusts—all show a delay in the onset of reproduction and a reduction in the number of young produced, as compared with non-migrating forms of the same species.

Left: the Bath white butterfly, a regular migrant through the passes of the Pyrenees; right: the voracious migratory locust is a threat to most of Africa

Restless activity and the urge to fly do not, in themselves, constitute migration, since they could result merely in trivial movements. Migration is characterized by the directed movements of the insects concerned, even though their final displacement may be determined mainly by the direction of the prevailing wind. As described in Exodus (*ii*. 12–19), an east wind brought the plague of locusts into Egypt, a west wind took them away and cast them into the Red Sea.

Both solitary grasshoppers and the comparatively few species, known as locusts, which periodically swarm gregariously in vast numbers, share the characteristic that their populations fluctuate greatly from year to year, depending upon the annual rainfall in the arid regions that these insects inhabit. In hot deserts, especially where irrigation provides permanent food for the developing stages, insect numbers can increase very rapidly. Every so often, vast swarms of Desert Locusts emerge from their scattered breeding grounds to ravage the crops of Africa and Asia. A single, medium-sized, swarm may contain over a thousand million insects, each eating its own weight every day, and may range over 4,800 kilometres (3,000 miles) from its starting-point. Although the area subject to invasions covers about 28,500,000 square kilometres (11 million square miles), only a fraction of this area is infected at any one time, even at the height of the plague, and breeding occurs only in certain areas where conditions are suitable.

The positions of the breeding sites vary from year to year because they are dependent upon rainfall that is erratic in distribution as well as in amount. Despite the fact that individual locusts steer by the sun, the net direction of movement of the swarms is downwind, so that most swarms are carried into regions of low pressure where rain has either fallen recently, or is about to fall. The significance of every locust flying in the same direction—whatever that may be in relation to the direction of the wind—is merely that it holds the swarm together. The advantage of this, in turn, is that individual insects are less vulnerable to the attentions of predators, which rapidly become satiated, than they would be if they were scattered over a larger area. Secondly, there is no difficulty in males and females finding each other for breeding.

The eggs are laid in damp sand, and by the time they hatch, a mass of tender grass shoots will have sprung up to provide food for the emerging hoppers. Displacement downwind, from an area in which the seasonal rains are ending to another where they are about to begin, seems to be essential for the maintenance of locust swarms when food supplies at the breeding sites decrease during the dry season.

When locust hatchlings have adequate living space, they remain in the 'solitary phase', developing, breeding and dying in their original homes in

exactly the same way as ordinary grasshoppers. When living conditions are crowded, however, either by the hatching of very large numbers, or by the contraction of the breeding grounds through climatic action, the hoppers become 'gregarious'. They develop a dark colour, quite different from the inconspicuous green of solitary individuals, and march together in bands often numbering many millions. For a few days after they have become adults, these locusts take short daily flights, which become increasingly longer as migration finally develops. Since 1910, there have been five widespread plagues of the Desert Locust (*Schistocerca gregaria*), and five major recessions.

From 1928 to 1941 swarms of Migratory Locusts (*Locusta migratoria*) invaded Africa, south of the Sahara. From their outbreak area in the flood plains of the River Niger, where they are now controlled, these locusts invaded some 17,223,000 square kilometres (6,650,000 square miles), much of which was subsequently infested by the Red Locust (*Nomadacris septemfasciata*), whose outbreak areas lie south of the Equator in Malawi and Tanzania. Much of Australia is liable to be affected in a similar way by the Plague Locust (*Chortoicetes terminifera*) and the small Plague Grasshopper (*Austroicetes cruciata*), while the Rocky Mountain Locust (*Melanoplus spretus*) sometimes swarms in the United States. Other species occasionally form migratory swarms in arid and semi-arid regions throughout the world.

The fact that migration takes place among dragonflies has been realized since 1673, when a mass flight was seen at Hildesheim, in Germany, proceeding from north to south. In 1857, a list was published summarizing nearly 50 dragonfly flights, about half of which had taken place in the Netherlands. Migration of dragonflies has since been recorded throughout the world. One common species regu-

Above: seven-spot ladybirds, one of the more important migratory species in Britain; right: a group of African migratory locusts. From their outbreak area in the upper reaches of the Niger River, swarms may infest almost the entire African continent

68

larly migrates up and down the Nile. Such migrations may be a source of danger to domestic fowls and turkeys, because the dragonflies transmit parasitic flukes, infection by which inhibits egg-laying and causes much ill-health among birds.

Long-distance migrations of ladybird beetles are not uncommon and, like those of the Bugong Moth, mentioned above, appear often to be journeys to and from hibernating or aestivating localities. In Britain, about half a dozen of the 40 species are known to show migratory movements and mass appearances. The most important of these are the Two-spot Ladybirds (*Coccinella bipunctata*), Seven-spot Ladybirds (*C. septempunctata*) and Eleven-spot Ladybirds (*C. undecimpunctata*). In 1847, huge swarms were swept ashore in Kent and in 1869 the streets of London were said to be swarming with Two-spot Ladybirds. On another occasion, about 4,500 million Eleven-spot Ladybirds were swept ashore on the north coast of Egypt, west of Alexandria, while millions more were still flying in the air. Ladybirds feed on aphids, and the movements of their migrating swarms are normally related to the highly mobile nature of the prey; sometimes they are blown off course by the wind.

The Convergent Ladybird (*Hippodamia convergens*) of North America feeds on greenfly in the coastal districts of California during the summer. When autumn comes, however, the beetles move towards the hills and congregate in masses, under stones and dead leaves, in pine woods, and sometimes even on open hillsides. Here they spend the winter in a state of hibernation, migrating back to the valleys when the spring arrives to feed on the aphids there.

At the beginning of this century, it occured to entomologists that it might be useful to collect Convergent Ladybirds during the winter, keep them in cold storage, and liberate them during the following summer whenever there appeared to be danger of an outbreak of greenfly. This was done on a large scale and for several years, but the experiment was a costly failure. The insects collected in their winter sleep had an instinct to migrate on awakening. Consequently, when they were released on the coast, most of them must have flown off into the Pacific Ocean, for they made no effect on the numbers of aphids.

The primary methods of locomotion in beetles are walking and swimming: the great majority of beetle flights are therefore migratory in nature. This is especially true of aquatic beetles, as of water-skaters and water-bugs; but certain leaf-eating beetles, including the notorious Colorado Potato Beetle (*Leptinotarsa decemlineata*) and the Spotted Cucumber Beetle (*Diabrotica duodecempunctata*), are also known to undertake migratory flights.

Migration occurs among sawflies, wasps, and flies, but, as flying is the normal method of loco-

Below: seven-spot ladybirds hibernating in autumn. They will migrate in the following spring

motion in such insects, it is difficult to distinguish between migratory flights and trivial movements. Sphegid wasps, however, are known to follow the swarms of locusts on which they prey. One of the earliest records of migration among two-winged flies is of myriads of hoverflies being found on the coast. It is now known that these insects, whose larvae prey on greenfly, undertake long migrations when their source of food has been exhausted. Hoverflies have been seen migrating through the mountain passes of the Pyrenees, and accompanying migrant butterflies through the Himalayan passes of Nepal. Many other mobile species, such as blow-flies and green-bottle flies, have been found to migrate, while fruit-flies, frit-flies and other smaller kinds form a part of the aerial plankton, as we have already seen, and are true migrants.

It is often impossible to distinguish, with certainty, between migratory and trivial movements in insects. If such movements take place at the beginning of adult life, however, and originate from a habitat that is deteriorating from the viewpoint of the insect, they are probably migratory—especially if the individuals engaged do not respond to food, mates or the environment above which they are flying. Migration takes many forms and is but one aspect of the chronological development of the behaviour patterns of insects. It is instinctive and may appear in the adult or larval stage.

Above: Colorado potato beetle, a species found in many parts of Europe as well as America, is known to undertake migratory flights of considerable length

71

Migration over Land

Apart from the fact that the distances covered are less, and speeds generally slower, terrestrial migration has remarkable parallels with aerial migration. Indeed terrestrial migration is occasionally seen in flightless birds. The ostrich (*Struthio camelus*) and Common Rhea (*Rhea americana*), for instance, wander nomadically over wide areas of desert and savanna. Although they do not appear to have any definite breeding grounds they often return during the dry season to the neighbourhood of water-holes or river beds.

Ground-living animals exhibit the whole range of migratory movements: trivial, nomadic, and irruptive, as well as true long-distance seasonal migrations; and migration is especially marked among the inhabitants of temporary habitats. For instance, most species of the tiny insects known as booklice live in permanent habitats beneath the bark of trees, but one kind that inhabits organic débris has occasionally been observed migrating in swarms.

Mammals

Whereas flying birds are the dominant aerial migrants, terrestrial mammals are the most important migratory animals on land. Some are nomadic, some irrupt in vast numbers, and some show true seasonal migration that is associated with nutrition and reproduction. Nomadism is characteristic of the inhabitants of arid and semi-arid lands. When the vegetation is too sparse to support a permanent population, the inhabitants have to travel far from one feeding ground to another. Where such nomadism is regulated by seasonal rainfall, it becomes a true migration, involving an outward and a return journey. Nomadic species frequently migrate to a special part of their range to give birth to the young, and since true migration is frequently correlated with reproduction, any distinction between migration and nomadism breaks down.

Nomadism is perhaps especially marked in mammals which occupy wide territories, such as elephants, bears, ungulates (hoofed mammals), and predatory carnivores that hunt in packs. In savanna regions, which experience seasonal drought, there are also considerable migrations of large mammals.

Broadly speaking, the large savanna mammals fall into three categories as regards their water requirements. The first category is water-dependent residents, such as hippopotamus (*Hippopotamus amphibius*), which are restricted to areas containing abundant permanent surface waters. This restriction does not, however, prevent hippos from making occasional long cross-country treks from one river to another, when under pressure from drought or local over-population.

The second category is arid-adapted resident species. These often show considerable physiological reduction in water-loss. They either use surface

Facing page: migrating herd of carabou in Alaska

Below: female ostrich feeding. Although flightless, ostriches wander nomadically over large areas of Africa

water for drinking or depend upon the water obtained from succulent or deep-rooted plants. Rhino are partially arid-adapted, non-migratory mammals. Their survival usually depends on the existence of thickets which contain succulent plants and also supply shade, so that the rhinos can lie up within them, thereby reducing water-loss to a minimum. In extreme conditions the rhinoceros also depends to a considerable extent upon the digging activities of African Elephants and other mammals to expose the sub-surface water in dry river beds and shaded wallows.

Water-dependent migratory or semi-migratory species make up the third category. Most important among the mammals in this group are African Elephants (*Loxodonta africana*), to a lesser degree buffalo, and to some extent predators such as lion (*Panthera leo*), cheetah (*Acinonyx jubatus*), Hunting Dogs (*Lycaon pictus*) and hyenas, as well as the insectivorous aardwolf (*Proteles cristatus*), the ratel or Honey-badger (*Mellivora capensis*), and the Bat-eared Fox (*Otocyon megalotis*).

Hoofed mammals do not normally have fixed territories, but spend their lives in constant movement: between migrations they are nomadic. When the breeding season approaches, however, the pregnant females of some species seek out temporary, secluded resting-places where they can safely give

Above, far left: the brown bear migrates only in exceptionally cold winters; left: hippopotamuses, seen here in Botswana, are restricted to regions with abundant surface waters; above: American bison wander over large areas following the grazing

Following pages: polar bears travel extensively, on land and water, and are also carried considerable distances by drifting ice floes

birth to their young. Even elephants do this, although the expectant mothers are often accompanied by another female from the herd who helps to protect the new-born calf.

Mobility varies greatly, not only between different species, but also between individuals of the same species. For example, a marked elephant in Zaire lived for several years consecutively within a radius of a few score kilometres, whereas large-scale and fairly regular movements have been observed among the elephants of East Africa. Again, the migrations of zebra and wildebeest (*Connochaetes taurinus*) are of spectacular dimensions on the Serengeti Plains, whereas, in South Africa, those animals are much less mobile. Clearly such differences are, to a large extent, related to seasonal changes in food supply.

It seems that migration to avoid climatic extremes is less common. Even in the islands of the Arctic Sea, neither the Musk-ox (*Ovibos moschatus*) nor the wolves (*Canis lupus*) that prey on the oxen move south in winter. The Arctic Fox (*Alopex lagopus*) even goes farther north at that time, following the Polar Bears (*Thalarctos maritimus*) and feeding on the remains of the seals that they kill. Lemmings and the Varying Hare (*Lepus timidus scoticus*) also remain during the winter, as do various other land-animals and a number of birds. The American Black Bear (*Ursus americanus*), too, does not move southward except during unusually cold winters, when it may migrate to an environment where it can

75

*Above: barren ground
caribou on migration;
left: caribou are unique
among deer in that
females may bear
antlers as well as
males; right: American
bison used to carry out
extensive migrations
before their movements
were restricted by
human activity*

hibernate without experiencing conditions so severe as to undermine its capacity for recovery.

In addition to local movements over a more or less restricted territory, some mammals make far longer seasonal journeys that may cover hundreds of kilometres and are associated with the breeding cycle. A classic example is the Barren Ground Caribou (*Rangifer tarandus*), which may travel 650–800 kilometres (400–500 miles) on their annual journeys. During summer, the caribou inhabit tundra but, from July onwards, begin a southward movement through the coniferous forests, along regular routes which are followed year after year. In some places the rocks are worn away to a depth of up to 60 centimetres (2 feet) by the thousands of caribou that have passed over them on countless annual migrations. Large herds are, in general, characteristic of the herbivorous animals of steppe land and savanna; sometimes the males move together in compact bands numbering between 100 and 1,000 head, but this segregation of the sexes is not invariable since copulation takes place on the autumn migration. The caribou remain in their winter quarters until the following spring when the northward journey begins again. The young are born at this time, but the caribou do not stop for long. They press onwards despite all obstacles, so that mass drownings sometimes occur when they attempt to cross flooding rivers. On one occasion

tions were of two types: localized wet-season nomadic wandering and long-distance, directional migrations covering many hundred kilometres each year. In different seasons, elephants invariably seek certain particular localities—open country during the rains, and forest in the dry season. Thus, near Kilimanjaro, elephants move down the northern slopes of the Usambara range in April and spread through the Nyika Plains almost to the coast. The wanderings of elephants in other parts of East Africa are likewise governed by the availability of food. Breeding herds, specially formed of animals from different herds, are said to have travelled from South Laikipia to the Aberdare Mountains, then north-east to the Lorian Swamps, north-west to Marsabit and then south again to the Aberdares. The round trip of some 650 kilometres (400 miles) takes three years. The young are born in Marsabit

as many as 525 corpses were found together.

When the American Bison was abundant, it used to carry out impressive migrations over a more or less circular route, the herds sometimes wintering up to 650 kilometres (400 miles) south of their summer territories. In contrast, the movements of the elk (*Cervus canadensis*) are much less extensive. They are comparable with the vertical migrations of Bighorn Sheep (*Ovis canadensis*), Mule Deer (*Odocoileus hemionus*) and moose which spend the summer feeding at higher altitudes on the mountains but, when the winter sets in, move to the relatively protected valleys where the snow is less deep and food more accessible.

At one time, African Elephants characteristically migrated over vast distances, so that they were assured adequate shade, nutritional variety, salt and water supplies throughout the year. Such migration also provided an opportunity for the herds to regroup, and large congregations, numbering over 100 animals, were sometimes found. These migra-

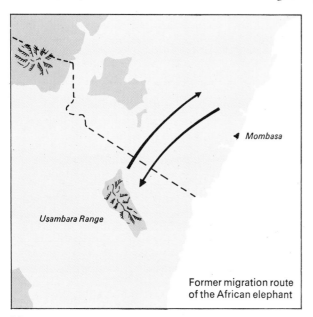

Former migration route
of the African elephant

Mombasa

Usambara Range

forest and the animals return with one-year-old calves.

Before human settlement blocked their progress, restricting them to game reserves and unpopulated regions, the game animals of eastern Africa moved with the seasons, crossing mountain ranges, wading through swamps and swimming rivers to reach the green savanna during the rainy season or to return to the forests with the onset of drought. In recent years, human settlement and agriculture has adversely affected the game by blocking migration routes and restricting the wanderings of animals to limited areas in which overgrazing and soil erosion frequently occur. The small pockets in which many species of African big game now survive may be relics lying on past migration routes.

Just over 480 kilometres (300 miles) south-east of Khartoum lies the beautiful, but little-known,

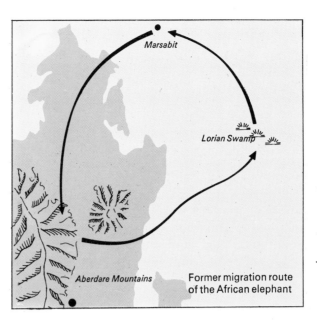

Marsabit

Lorian Swamp

Aberdare Mountains

Former migration route of the African elephant

Far left: Grant's zebra and young. Zebras migrate regularly throughout their range as they follow the fresh grass that springs up after the rains; below: elephants moving in the plains beside Kilimanjaro; maps at far bottom left and left show former migration routes of the African elephant

Above: Grant's zebra and young. Zebras often migrate in company with antelope and wildebeest; far right: wildebeest drinking from the Seronera River, Serengeti; right: migration route of the wildebeest in the Serengeti region

Ethiopia and the Sudan. The migration begins during the month of May, when the swamps of the Upper Nile rise, and there is a general movement of animals south-east towards the arid Kenya border. The herds of antelopes, thousands strong, sound like massed cavalry as they approach over the bare horizon. The predominent species are White-eared Kob (*Adenota kob*), Tiang (*Damaliscus korrigum*) and Mongalla Gazelle (*Gazella albonotata*). Zebra, Grant's Gazelle (*Gazella granti*), Lesser Eland (*Taurotragus oryx*) and buffalo (*Syncerus caffer*) also move in considerable numbers. Some oryx (*Oryx gazella*) and smaller numbers of giraffe (*Giraffa camelopardalis*), waterbuck and Roan Antelope go along with the main body which is flanked by lion and the smaller predators.

Until a few years ago, the plains of southern Ethiopia and northern Kenya were thickly populated every year at the end of June, not by hundreds, but by thousands, of the species of game that customarily undertook the hazardous migration to the south. Here they were checked by the Turkana Desert plains, but remained contentedly for three or four months until the need for fresh grazing drove them north again in the fertile tracks of the retiring rains. By September, the area would begin to clear again. In massed columns many kilometres long, the game moved slowly and benignly northwards, shielding its young from voracious predators, and the plains were given back to the burning heat. Herds of oryx and Grant's Gazelle, scattered into the outlying stretches by the migration, returned now to the forsaken landscape. One could stroll through country whose every square kilometre held its complement of many hundred kob, and then abruptly, for no apparent reason, an invisible boundary was crossed and one would see none. One such boundary lay east of the Loelli landing-strip. At the height of the migration this strip held as many as 3,000 kob; and yet a few hundred metres

Dinder National Park. To get there, you drive past blue irrigation canals along endless corrugated desert tracks, crossing the Blue Nile over the Sennar Dam. Suddenly the desert acacias give place to *talih* with vultures, bee-eaters, Red-hussar (*Erythrocebus patas*), and Grivet Monkeys (*Cercopithecus aethiops*) in their branches, and graceful gazelle crossing the track beneath. All around, the Common Guinea-fowl (*Numida meleagris*) can be heard calling among the trees. In the *Beit el-Wahash* (home of beasts) there are vast herds of delicate reedbuck (*Redunca arundinum*), Bushbuck (*Tragelaphus scriptus*), sombre, stocky waterbuck (*Kobus defassa*) with massive horns, Roan Antelope (*Hippotragus equinus*), warthogs (*Phacochoerus aethiopicus*) and ostrich that stretch like armoured divisions to the hazy horizon.

Such a mixing of quite unrelated species comes initially as something of a surprise to the ecologist who is accustomed to the idea that each species occupies the particular habitat to which it is specialized. The explanation is that the different species are indeed separate throughout much of their lives, but that they come together at certain times at places such as water-holes, salt-licks and the like. Those that migrate to the Dinder River during the dry season each year move back across the border into Ethiopia during the rains.

One immense, but little-known, game migration takes place each year between Kenya, south-west

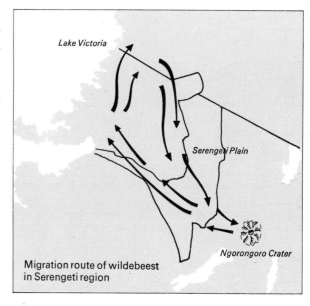

Lake Victoria

Serengeti Plain

Ngorongoro Crater

Migration route of wildebeest in Serengeti region

east, one might search for days and not come across a single animal.

When the dry season begins in June or July, the wildebeest of the Serengeti move in thousands on a 320-kilometre (200-mile) journey westwards towards Lake Victoria, returning when the rains have revived the parched grasslands. Here one may still see some of the remaining great herds of African plains game animals and their attendent predators. On the one hand the wildebeest, zebras, buffalo and numerous species of antelope; on the other lions, leopards (*Panthera pardus*), cheetahs, hyenas, Hunting Dogs and jackals. Most of these plains game are migratory animals and are confined to arid regions with seasonal and uncertain rainfall. In order to survive, they have to migrate between their areas of wet-season and dry-season grazing. The former lie on the Serengeti Plains in the lee of the Crater Highlands; the latter could be in the Highlands themselves, only about 32 milometres (20 miles) away.

Unfortunately the animals follow a very much longer migration cycle which takes them outside the borders of the Serengeti National Park for a large part of every year. They follow the river valleys running from the plateau on which the plains lie, down to Lake Victoria where there are alluvial pastures called *dambos* or *mbugas*. These are water-logged during the rains so that there is little tree

Right: African buffalo with cattle egret. Buffaloes often migrate in large herds

Below: hyaenas feeding on a dead elephant calf. Carnivores follow the migrating herds of herbivores in Africa

growth, but they provide good pasture during the dry season. The intervening migration routes are mostly through 'bush' country of various types, often heavily infested by tsetse.

On the Serengeti Plains, the game compete for pasture and water with the herds of the Masai, a tribe whose men live on blood and milk from their cattle. The Masai have similar migratory habits to the game, and their steady destruction of the highland forests almost certainly has an adverse effect on the water régime. Survival of the plains game therefore depends on sacrificing to them a share of both wet- and of dry-season grazing and probably also on maintaining the fertility of the intervening bush.

Hippos tend to migrate up river when the higher reaches are full and return to the mouth in the dry season. Rhinos, on the other hand, usually do not migrate. Nor do lions and other predators, except in pursuit of their prey; they change their hunting grounds according to the supply of game. Kudu and other antelope regularly migrate to summer feeding grounds, where they calve, returning to more favourable quarters for the winter. The quaggas (*Equus quagga*) of South Africa were alleged to migrate regularly in bands of two or three hundred.

The most dramatic example of migration among African game is afforded by the springbucks (*Antidorcas marsupialis*) of south-east Africa which, at one time, would migrate on occasion in unbelievably vast numbers. On the larger migrations, troops of 10,000–20,000 gathered together into columns numbering hundreds of millions. During 1849, the town of Beaufort West was invaded by a vast number of Springbuck, accompanied by Blesbok (*Damaliscus dorcas*), quaggas, wildebeest, eland and antelopes of all sorts and kinds, which filled the streets and gardens as far as could be seen. After three days the horde disappeared, leaving the country looking as if a fire had passed over. Another vast emigration has also been described in which a wide column took several days to pass the same spot, though it was not always of the same density. Many died, especially old animals and kids, but the survivors headed onwards until they reached the sea where they were drowned in such numbers that for nearly 50 kilometres (30 miles) the beach was piled high with corpses. The migration was probably due to food shortage after a population build-up.

Other big treks of springbucks have been recorded in which the animals streamed past like the floods of great rivers, covering the plains and hillsides in vast masses. On one great emigration in Namaqualand, the herd reached the Atlantic Ocean, where they dashed into the waves, drank the salt water and died.

Lions, leopards, Hunting Dogs, hyenas, jackals, as well as other carnivorous mammals and birds of prey used to follow the migrating herds, feeding on them when they were so inclined. So prolific were the springbuck, however, that their numbers were not depleted until the advent of the European colonists.

The Boers divided springbucks into two categories: *hou-bokkers*, which remained permanently in particular areas of the veldt; and *trek-bokkers*, which occupied less favourable country and were subject to the vast emigrations mentioned above. This suggests that food-shortage, coupled with an excess population increase, rather than drought, may have been the driving factor in springbuck migration, especially as the springbuck in the Kalahari Desert can live without water on dry food and succulent roots.

During their treks, the normal behaviour of the springbuck changes considerably. The animals become restless, wander aimlessly, are startled without cause and gallop off in any direction until they collect together again. Yet they lose their natural shyness and even enter villages and towns. During the great springbuck treks of the last century, some of the animals in search of water were drinking from fountains in the streets of townships. Changes of behaviour during migration, however, are a common occurrence in other kinds of animals and no real explanation of the phenomenon has yet been given. The last springbuck migration took place in South-West Africa in 1954. It was on a comparatively small scale.

The mass hysteria and lack of shyness among emigrating springbuck, caribou, lemmings and other animals may be associated with hormonal imbalance due to overcrowding or food shortage. Overgrazing, disease, or stress may also be contributory factors in some cases. Although the migrating springbuck hordes, for instance, were sometimes fat, they were more often in poor condition and sometimes heavily infested with scab. The *trek-bokkers* of 1896 did not disperse to the Orange River, but migrated into the dry interior where the does kidded. It is not unusual for migrating animals to give birth during the course of their journeys, but this and other examples do suggest that the dispersal may have been in search of forage and that hysteria may be the result of food shortage.

Few animal populations remain constant in size for very long. In addition to seasonal changes in numbers, cyclic variations in population size over periods of several years are seen, especially among the birds and mammals of the Arctic and cool temperature regions of the globe. Two or three of these cycles seem to occur. First, there is a three-year or four-year cycle in lemmings and the animals that feed on them. When lemming populations crash, the Snowy Owl is forced by starvation to migrate for hundreds of kilometres to the south. Secondly, there is a four-year cycle in the number of animals inhabiting the belt of open forest lying between the

Far right: lemmings are small voles of Northern Europe, Asia and America. They emigrate in vast hordes when population numbers reach a peak; below: the nomadic springbuck of southern Africa in company with a herd of wildebeest

tundra and the coniferous forest or *taiga*. This is based on voles. Thirdly, there is a ten-year cycle in populations of the Snowshoe Rabbit (*Lepus americanus*) and other animals of the northern forest regions of North America. When population numbers reach their peaks, irruptions sometimes occur. The regularity of cycles in population numbers in the Arctic may be due to the fact that basic food chains are comparatively simple there, and predator-prey oscillations are little disturbed by other factors.

Better known, though perhaps less spectacular than the migrations of springbuck, are the irruptions of lemmings. Lemmings are represented by different species in Europe, Asia and North America. They form the basic food of the predatory mammals and birds of the Arctic regions, and their irruptions are followed by many of these enemies—in particular, by ermine, wolves, Arctic Foxes, falcons, buzzards, skuas and Snowy Owls. In certain years, the numbers of offspring in the lemmings' litters are more than usual, and the young themselves mature and reproduce sooner than usual, so that the population reaches a maximum at the end of summer when food becomes scarce. It is at times such as these that the animals at the outskirts of the congested areas begin to forage further afield and are followed by other emigrants, crowded out from the regions of greatest famine. Most of the emigrants

are young males which have been unsuccessful in establishing territories for themselves.

Once emigration has begun, lemmings, like springbuck, lose their normal shyness and do not shrink from attracting attention. They seem to have no fear as they push irresistibly down the rough mountain sides, tumbling down slopes and into holes regardless of whether they can climb out again, or not. They enter towns, even invading the houses, and falling prey there to cats and dogs. They fight each other and anything that opposes them, although being half-starved they stand little chance.

The spectacular nature of lemming migrations is partly due to the nature of the country. In Norway, for example, mountain lemmings inhabit high fells on the slopes and plateaux above the tree line, and each area is isolated by lofty peaks or steep valleys. The tide of emigration is therefore restricted and flows down to the lowlands through certain passes in swarms large enough to attract attention. The smaller Wood Lemmings (*Myopus schisticolor*) also increase in numbers and migrate in the forests, but their movements are less conspicuous and the animals quickly arrive at new feeding grounds.

Lemmings from Greenland migrate over the frozen ocean to the mainland from islands up to 80 kilometres (50 miles) away. When the ice breaks, the little creatures run backwards and forwards

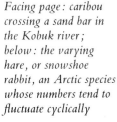

Facing page: caribou crossing a sand bar in the Kobuk river; below: the varying hare, or snowshoe rabbit, an Arctic species whose numbers tend to fluctuate cyclically

alone the banks of streams and rivers, looking for a smooth place with a slow current where they may safely cross. Having found one, they at once jump in and swim fast to the other side where they give themselves a good shake, as a dog would, and then continue their journey as though nothing had happened. Lemmings move erratically and no definite path is chosen, although the general direction is partly determined by the nature of the ground across which they are travelling. Water is no obstacle, as already mentioned, and lemmings have been known to swim across fiords more than four kilometres (2½ miles) wide. It is not surprising, therefore, that they should sometimes be swept away by the tide: there is no truth in the story that lemmings march in an army to the coast and commit suicide in the water.

This legend of self-destruction may have arisen from disasters in which large numbers of lemmings have inadvertently perished at sea. In 1868, for instance, a steamer in Trondjem Fiord, sailed for 15 minutes through a herd of swimming and drowning lemmings, and similar observations have been made subsequently. Migrating hares and rabbits in the Arctic are also known to swim across rivers. In 1867, a swarm of squirrels was reported to have reached Tabolsk in the Urals, crossing the Tehussoveia river by climbing the oars of boats.

Myriapods

Since they feed on decaying plant material, wood-lice and millipedes might well be expected to undergo irruptions and mass migrations as, indeed, they do. Oddly enough, however, the earliest account (cBC 220) of migration in myriapods (literally many-footed) refers to neither of these but to centipedes. It is found in Aelian (*xv.* 26) and here it is cited in Charles Owen's *An Essay towards the Natural History of Serpents*, published in 1752:

> These little creatures (centipedes) make but a mean figure in the ranks of animals, yet have been terrible in their exploits, particularly in driving people out of their country. Thus the people of Rhytium, a town in Crete, were constrained to leave their quarters for them.

The earliest recorded migration of millipedes took place in March and April 1876, when great numbers of a variety of species (accompanied also by centipedes) migrated in Transylvania. Two years later an enormous mass of millipedes actually stopped a train on the Thein Railway in the Hungarian district of Alföld. The millipedes were in such vast numbers that the earth was black with them. The locomotives destroyed them in thousands on the rails and were impeded to such an extent that sand had to be strewed before their driving wheels would grip.

Trains were stopped in a similar manner, but by a different species of millipede (*Schizophyllum sabulosum*) in northern France in 1900, in a wood on the railway from Lutterback. According to the stationmaster at Sennheim, a goods train came to a halt at 6.35pm on 5 June because the track was made slippery by numbers of squashed millipedes, small parties of which were crossing from one side of the line to another at intervals over a distance of a kilo-metre. Had the train being going down hill they would, in all probability, not have been noticed. In fact, a passenger train had shortly before passed in the opposite direction and, presumably, squashed a lot. Similar migrations were recorded in Prussia in 1906 and 1938. Only one mass migration of millipedes has been recorded from Britain—in May 1885, large numbers of *Tachypodoiulus niger* were seen crossing a road from a field of oats to one of pasture—but others have been reported from Yugoslavia, Rumania, Sweden, Poland, and in Latvia. In some instances, the migrating millipedes were accompanied by woodlice.

Many more instances of mass migration of milli-pedes have been recorded from North America, and the majority of these took place in West Virginia. In 1818, a great army covered over 30 hectares (75 acres) of farmland at Littleton in central West Virginia. Four weeks previously, they had been seen moving slowly south-west from a wood a couple of

90

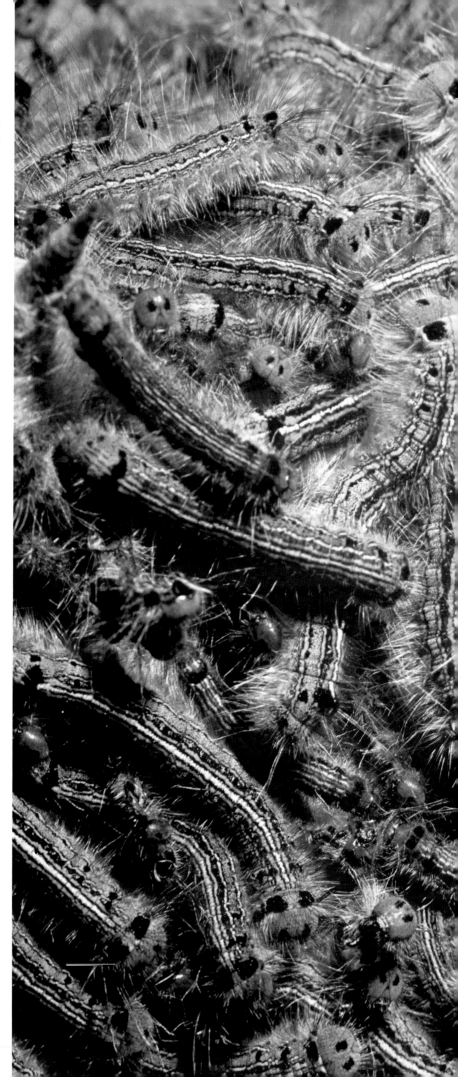

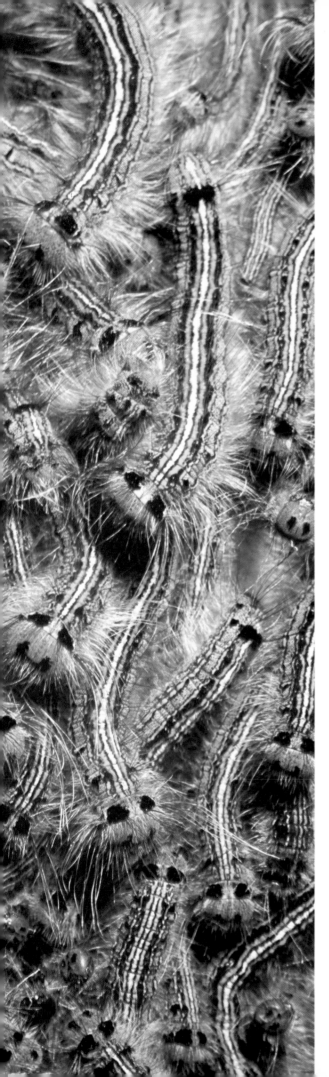

kilometres away. The millipides were so numerous that cattle refused to graze on the invaded pasture, wells were filled with drowned corpses to a depth of 15–20 centimetres (6–8 inches), and workmen cultivating the corn became nauseated and dizzy from the odour of millipedes crushed by their hoes. On warm, clear days, masses half as large as a barrel would collect in damp and shady places, but in cloudy weather and at night, the army was constantly on the move. Rotting wood and old stumps were gnawed white in many places where the animals had passed, and covered with small dots of earthy excrement. Eventually the majority, estimated at 65,340,000 millipedes, perished at the bottom of a cliff, killed by the hot sunshine. Nothing fed on them.

Similar companies of different species have been seen in subsequent years in Virginia, Louisiana, Ohio, California, Arizona, Texas, and New Mexico. Mass migrations of large tropical millipedes have been reported in East and West Africa, as well as in South America. Although certain aspects of the phenomenon are still not explained, it seems to follow abnormal reproductive success and consequent overcrowding, and to be engendered by drought or stimulated by sudden drops in temperature.

The larvae of certain moths, known as 'processionary caterpillars', are gregarious and construct tents of silk in which they take refuge from inclement weather. They feed at night, travelling to and from the nest in a single-file procession, each larva spinning a thread of silk wherever it goes. The threads of silk left behind by a big procession become

Left: brightly coloured caterpillars of the lackey moth are gregarious and may migrate when too numerous; below: migrating millipedes near Lake Chad

quite thick, in some places forming a band two or three millimetres (an eighth of an inch) wide. But it is not the threads that guide the caterpillars so much as the tails of the larvae in front.

The same phenomenon of procession is found among the larvae of certain saw-flies. No doubt the conspicuous mass movements of these distasteful, brightly coloured insects warn off potential predators which might attack individuals but avoid such swarms. The larvae of one of the fungus-gnats provide a similar example. When fully grown and fed, large numbers of these assemble and travel together in long columns.

Army-worms and snake-worms are well known agricultural pests. They are the larvae of a number of moth or butterfly species that become migratory when crowded, as locusts do. The migratory phase is darker in colour, more active, and has a higher fat content than the solitary phase. Migratory caterpillars are intensely gregarious in their mass-emigration, and mostly belong to species that are well-known as migrants when they become adult. The Silver Y-Moth (*Plusia gamma*), for example, is one of the commonest immigrants to the countries of northern Europe. In some summers, these moths appear in incalculable numbers and reproduce, but they do not overwinter in the higher latitudes. Their permanent home is in the arid belt extending from the shores of the Mediterranean eastwards into Asia. Similarly, American army-worms also emigrate in the adult state. However the majority of insects that are especially notable for the irruptive swarms they produce have not evolved larval-phase differences of this kind. Not surprisingly, migration by walking over land plays a much smaller part in the distribution of insects than does dispersal by flight.

Amphibians and reptiles

The migratory instinct is well marked in frogs, toads, newts and salamanders, which may travel considerable distances in order to collect together in a special pool or stream for purposes of reproduction. Frogs are known to return to the same pond year after year in the breeding season. Though the owners may plant laburnum, hawthorn and flowering shrubs, though their children learn to skate there, and old sportsmen enjoy curling matches in the winter, these are mere diversions. Every spring the frogs assert their rights and arrive in their hundreds. And every summer thousands of tadpoles delight the youth of the neighbourhood, and innumerable tiny frogs, clean and shining, leave the water to invade the surrounding countryside. Many leave, but few, alas, return; the life of a small frog qualifies for no insurance. Furthermore, so many ponds have been filled up and ploughed over during recent years, that the Common Frog (*Rana temporaria*) is rapidly becoming an endangered species

Far left: Army worm moth larvae in West Africa where they defoliate grass and are a serious pest in cattle-raising regions; left: The silver Y-moth, a common immigrant into Northern Europe

Below: The European fire salamander may travel a considerable distance to the water where it breeds

except perhaps in suburban gardens where it can reproduce unmolested in ornamental ponds. It is not known for certain how frogs find their way to a pond. Some say that the smell of water attracts them, but this has not yet been proved.

In desert regions, heavy rainfall may stimulate the breeding of toads so that, when the temporary rain pools dry up, the barren sands are infested by numbers of baby toads emigrating in search of some relatively damp, cool shelter where they can hide away until the drought is again broken by a welcome deluge.

Because their skin is porous, and evaporation takes place rather quickly, most amphibians move about at night, when the humidity is greater than it is during the day. The periodic migrations and breeding of many frogs and salamanders are initiated by rain, although temperature may also be an

important factor. In temperate regions of the world, land-living amphibians migrate to water to breed, the males usually preceding the females by a day or two. This order of arrival at the breeding grounds is also found in purely aquatic species, and is characteristic of many other groups of vertebrates. As with birds, the male frogs and toads immediately establish territories and calling-stations, and then endeavour to attract females towards them by strenuous croaking.

Toads and frogs usually migrate to ponds to breed, some salamanders choose mountain streams, while certain tropical tree-frogs descend to the ground or from trees to bushes overhanging water. Amphibians have been shown to be attracted by the smell of moist soil and decayed wood, and they also respond to gravity. It is probable that the individuals which return to exactly the same breeding site for several years in succession, may be reacting to visual impressions and, perhaps, to a memory of distances travelled.

Preserving the eggs

Emigration from the breeding area by the tiny young is a hazardous process and many of them become the victims of birds and other predatory enemies. It is not surprising that so many eggs should be laid. Many of the amphibians of tropical rain-forests, however, carry their eggs and tadpoles around with them. The Surinam Toad (*Pipa americana*) is almost entirely aquatic, but does not spawn into the water. Instead, the eggs are glued to the back of the female when they fall into invaginations (tubular sheaths) of the skin. As development proceeds, each egg becomes concealed in a pouch or cavity with a lid. Still other frogs carry their eggs and tadpoles around until the young emerge as miniature adults. Such traits have been attributed to the rarity of permanent lakes and ponds in the rain-forest, and of standing water generally—apart from the rivers and their permanent streams which are populated by a rich, but dangerous, fauna of predatory water-beetles, water-bugs and dragonfly larvae. Whatever their function, these habits must have survival value and eliminate the need for migration to more secure areas.

Migration is an important feature in the lives of turtles and other marine reptiles, as the following chapter shows, but its importance to terrestrial forms is not great. Emigrations of large numbers of crocodiles have, however, occasionally been recorded and seasonal movements have been described in the viper in southern Finland. During the mating season, which lasts for about a month, the males move about nomadically in search of females. Early in June, they leave the wintering areas for their summer territories—which may be a couple of kilometres (over a mile) away—returning there in mid-September. Throughout the long winter, they hibernate in dens among boulders and under hillocks and tree stumps, often in company with frogs, toads, lizards, slow-worms, and grass snakes. Similar small-scale seasonal movements of land-reptiles take place in other parts of the world, but they are probably of little significance in the lives of individuals, though they may assist in dispersal of the species.

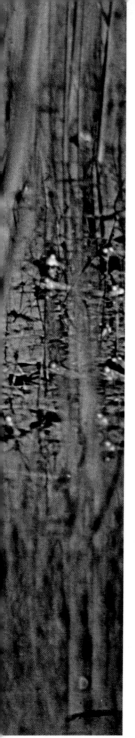

Above: migrating frogs travel long distances to congregate in a particular stretch of water, to which some return year after year; right: European tree frog. Most tree frogs are climbers and enjoy an arboreal life. This species, widely dispersed throughout Eurasia, breeds in ponds and streams

Migration in Water

For migratory animals, water is a medium more similar in some ways to air than to land. The oceans and seas of the world, all of which are inhabitable, have not only an immense horizontal area, but also considerable depth, whose average may be as much as three kilometres (about 2 miles). Marine animals, therefore, are free to move great distances, and the vertical movements of drifting, or planktonic, animals, are as much part of migration in water as the various types of horizontal migrations that occur.

Vertical migration of plankton

Vertical daily movements are a characteristic feature of the behaviour of many free-living planktonic plants and animals. The smallest planktonic organisms are bacteria and diatoms, but the floating flora and fauna of the sea also include innumerable blue-green algae and flagellates. Upon these depends the animal plankton, or zooplankton, which includes protozoans, small jellyfishes, larval worms, star-fishes and sea-urchins, molluscs, fishes and, above all, crustaceans—both larval and adult. This vast diversity of microscopic animals, in turn, provides the basic food of the fishes which are themselves eaten by larger fishes, predatory squids, seals, whales and sea-birds. The whalebone whales also feed on planktonic crustaceans or 'krill' which they filter from the water with their plates of baleen.

The migration of zooplankton at night, from a depth of at least 200 metres (660 feet) towards the surface of the sea, has been known since 1872, when HMS *Challenger* was despatched on the famous expedition during which she sailed for three and a half years exploring the oceans of the world. Since that time, the widespread nature of this phenomenon has been recognized as one of the most striking and universal aspects of planktonic life. Nevertheless, it is by no means properly understood, even today.

In general, most planktonic species avoid strong light, each showing a preference for a certain intensity. For this reason few organisms are to be found in the surface layers of the sea during the hours of daylight. They are distributed at various depths according to their specific light response. At dusk

they tend to swim upwards, but when all is dark and there is no light stimulus, planktonic forms tend to scatter. They migrate to the surface again at daybreak and later move downwards as the light strengthens. Doubtless the plants, or phytoplankton, tend to aggregate in the light intensity best suited for photosynthesis, just as the animals tend to remain in regions where they find a rich supply of food. This is not the whole explanation, however. There is often an inverse relationship between the distribution of the plant and animal plankton.

Diurnal migrations occur even among deep-water animals, and the movements of the species most abundant far down in the Sargasso Sea, for example, are obviously related to day and night. Moreover,

Facing page a common porpoise slicing through the waters of the Atlantic ocean. These animals migrate as far north as Greenland, Spitzbergen and the White Sea; below: phyllosoma larva of a Mediterranean lobster. Crustacean larvae form an important element of the zooplankton and undergo daily vertical movements in the sea

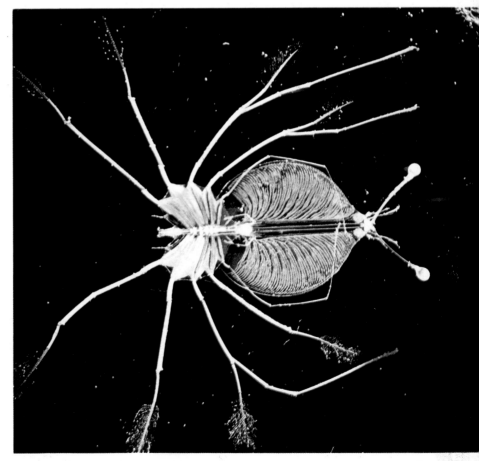

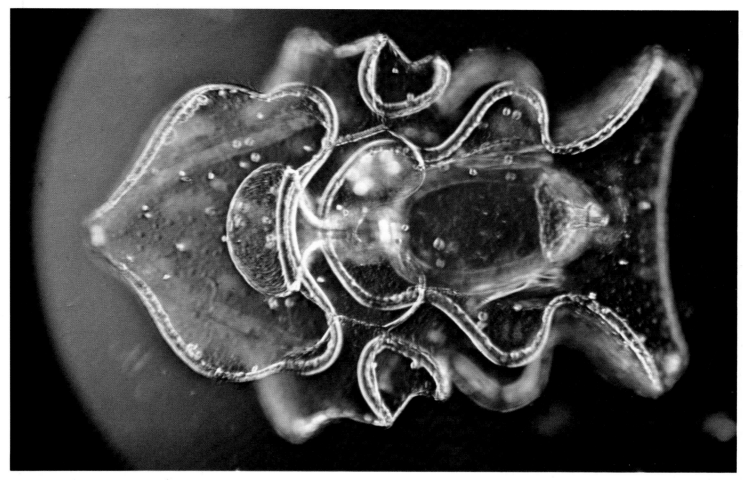

the limit of light penetration, about 1,000 metres (3,300 feet) in clear water is not the only factor determining the vertical distribution of animals. Other important considerations are marked changes in temperature, salinity, oxygen, and phosphate content. It has therefore been suggested that a combination of factors may be responsible for keeping the majority of the larger planktonic animals above 1,000 metres in these waters.

The various factors of the environment controlling diurnal migration of planktonic animals have been listed in the following order of importance: light, which clearly dominates under average conditions; temperature, which becomes very important and can even overwhelm the effect of light when it exceeds 20°C (68°F); and, finally, other factors such as salinity and aeration. The actual mechanism by which organisms keep themselves at the optimum level may be responses to light, gravity and the acceleration or inhibition of movement. It may be a combination of them all, and from the evidence of laboratory experiments, would seem to vary in different species.

The diurnal cycle of vertical migration of planktonic crustaceans has been shown to consist of four parts: ascent from the day depth, midnight sinking, dawn rise, and descent again to the daytime depth. Ascent in the evening and descent in the morning are related to decreasing and increasing penetration of

daylight. The midnight sinking is probably the result of the passive state of the organisms in full darkness, whereas the dawn rise represents a return by the animals to optimum light intensity. This is supported by the fact that the order in time of arrival at the surface for some fresh-water species is the same as the order in depth at which they are to be found in full daylight.

It has recently been shown that a population of water-fleas in a tank filled with a suspension of india ink in tap water will undergo a complete cycle of vertical migration. A 'dawn rise' to the surface at low light intensity is followed by a descent of the animals to a characteristic maximum depth. The animals rise to the surface again as the light decreases and finally show a typical midnight sinking. The light intensities at the level of the animals are of the same order as those observed in field observations.

Optimum light intensity may be an important factor in influencing the vertical migration of planktonic organisms, but a number of other factors are also involved. Some animals which usually stay down in the daytime may occasionally be seen at the surface in bright sunshine. And it is clear from the results of tow-netting at different levels that all individuals in a population do not react in the same way to one particular set of conditions. Even though the majority may migrate upwards, a proportion usually remains below.

The auricularia larva of a sea-cucumber. These ciliated larvae are prominent members of the marine zooplankton

One theory is that vertical migration may have been evolved because it gives the animal concerned a continual change of environment which would otherwise be unattainable for a passively drifting creature. Water masses hardly ever move at the same speed at different depths, for the surface areas are nearly always travelling faster than the lower layers. By swimming in a horizontal direction only, an animal would not get much change of environment in the sea, but by moving upwards and downwards, it can achieve an extensive degree of movement. Vertical diurnal migration of plankton is found the world over and clearly has important adaptive functions, but the exact nature and relative importance of these has yet to be assessed.

Crabs and fishes

Unlike the diurnal movements of planktonic crustacean larvae, the rarer seasonal movements of larger crustaceans, such as lobsters and crabs, are concerned with reproduction. Thus, the edible crab migrates into deeper waters in the autumn, and spawning occurs during the winter. Early in spring, however, the berried female crabs, carrying their developing eggs attached to their bodies, return to shallow water. Here the young hatch. In late summer, the females moult and are fertilized by the males which do not migrate. Then they return to deep water again for the winter. Numerous similar breeding migrations are known among other species of crabs and lobsters, while land-crabs annually

Close-up of a spiny lobster. Sometimes these crustaceans wander at random, while at others they migrate in large numbers

migrate in mass to the sea to reproduce. So, too, do land hermit crabs (*Coenobita* spp.) and the giant Robber Crabs (*Birgus latra*) of tropical coral islands, which feed on the pulp of coconuts.

Observations on tagged or marked individuals of the Florida Spiny Lobster (*Panulirus argas*) have indicated that these animals tend to wander at random, since over 90 per cent of a total of 251 recovered were not within more than 30 kilometres (19 miles) from the site of release and there was no evidence of seasonal migration along the Florida coast. In contrast, massive migratory movements of the same species have been recorded off Bimini in the Bahamas, the lobsters heading southwards in long queues—but neither their origin nor their destination is known.

Although any given species of fish will have a definite geographical distribution, shoals are constantly on the move within these areas, and may undertake extensive migrations. These are nearly always concerned with nutrition or reproduction. For instance, huge schools of tunny (*Thunnus* spp.) enter the Mediterranean early in summer following the shoals of smaller fishes on which they feed. Mackerel (*Scomber scombus*) keep to open water during winter, but in summer, when the seas are warmer, they approach the coasts on both sides of the North Atlantic. After spawning on the Continental Shelf, they move into bays and estuaries where they find an abundance of young fish fry to eat. The pilchard (*Sardina pilchardus*) also enjoys warm water, and retires to it during winter; but in summer, it approaches the coast of Cornwall, the northernmost limit of its range. The anchovy (*Engraulis encrasicholus*) requires warmth too and passes up the English Channel in spring to spawn in the estuary of the River Scheld, followed by its predator, the Horse Mackerel (*Trachurus trachurus*).

For many years, the seasonal movements of herrings (*Clupea harengus*) have attracted the interest of scientists because of the great economic importance of these fishes. The wealth of the Hanseatic League was founded on the herring fisheries of the Baltic, and it declined when the herrings moved

A school of pilchard. These gregarious fishes migrate to warmer waters in winter

elsewhere. At some seasons, herrings may be found in huge numbers in a particular region, then they disappear entirely. The movements of their shoals are not fully understood, but the species is evidently divided into a number of separate populations, each with its own range of distribution and its own spawning season. Herrings collect together both as young fishes and, when adult, before, during, and after spawning.

Marine fishes, which have definite territories to which they regularly migrate, include the barracuda or 'Snoek' (*Sphyraena barracuda*). This spawns in the coastal waters of southern Africa and then returns to deep water during the winter season. Others that migrate to inshore waters during the breeding season include blennies, hake (*Merluccius merluccius*), sprat (*Clupea sprattus*), sea bream, Red Mullet (*Mullus surmuletus*) and sharks. Perhaps the most remarkable example of migration into shallow water for breeding is the grunions (*Leuresthes tenuis*) of the North Pacific. These small fishes migrate in breeding shoals of many thousands to the sandy shores of the coast of California. Here they swim right through the surf and are washed ashore by the waves, near the high-water mark during the 'spring' tides of March, April, May and June. Almost on dry land, and above the level of normal tides, the male and female burrow together, the eggs are laid and fertilized, and the parents wriggle back into

Above: a school of young herring on migration; left: mackerel keep to open water during winter but approach the Atlantic coasts when the seas are warmer

101

the sea. Ten days later, when the tides are again exceptionally high, the eggs are on the point of hatching. As they are washed out of the sand by the waves, the young emerge and escape to the sea in the wash of the retreating wave.

After spawning in 'nursery grounds' on the eastern side of the North Sea, cod (*Gadus morhua*) disperse in various directions. Some move inshore to the rocky coast, some seek rough ground north of the Dogger Bank, while others make for the area where the herrings are found on the western side of the North Sea. In all cases the plentiful food they obtain enables the cod to recover quickly from the spent condition in which they find themselves after discharging their abundant eggs and spermatozoa.

Many fish that feed and mature in the sea ascend rivers when their reproductive instinct grows strong, and breed in fresh water. Examples include the Pacific Salmon (*Oncorhynchus* spp.), the Atlantic Salmon (*Salmo salar*), sturgeon (*Acipenser* spp.), shad (*Alosa* spp.) and the killifish (*Fundulus heteroditus*). These are known as 'anadromous' fishes, in contrast to 'catadromous' species who descend to the sea for breeding from their usual habitat of rivers and lakes.

The breeding season of the Atlantic Salmon takes place mainly in November and December, although the fish may approach the coast and enter suitable rivers in almost every month of the year. At first, they are fat and silvery, but as the time of spawning approaches, their colour changes to a dull, reddish brown, the front teeth of the males become enlarged, the snout and lower jaw are drawn out, and the lower jaw is turned upwards at the tip to form a prominent hook or 'kype'. In addition, the skin of the back grows thick and spongy so that the scales become embedded in it, and red, orange and large black spots make their appearance. Ripe females are darker in colour. After spawning, the spent fish, which are now known as 'kelts' are in a very enfeebled condition and many succumb to disease or are eaten by predators. Those that reach the sea, however, are soon restored to their normal silvery condition, remaining so until they again migrate to fresh water to reproduce. Salmon do not necessarily reproduce every year, and it is rare for an individual to spawn more than three times during a life of eight or nine years.

Pacific Salmon, of which there are six species, including the Sock-eye Salmon (*Oncorhynchus nerka*), Humpback Salmon (*O. gorbuscha*) and King Salmon (*O. tshawtyscha*), may migrate 3,200 kilometres (2,000 miles) from the sea to their breeding grounds inland. They have somewhat similar life-cycles to that of the Atlantic Salmon, although the differences between the sexes in the breeding season are even more marked. The reproductive act seems to exhaust them even more, for, after spawning, the spent fish drift helplessly downstream and none succeed in

102

Migration routes of the Downs herring

Migration routes of the Buchan herring

Left: migration of the Downs herring population; below left: migration of the Buchan herring. The range of herring distribution and period of spawning varies with different populations

reaching the sea alive. In some Canadian rivers, the corpses of dead salmon may be seen lining the banks for miles, sometimes piled up to a height of a metre or more.

The change in salinity experienced during migrations from fresh water to the sea and *vice versa* impose severe physiological problems on fishes. There is a tendency for fresh water to pass into the body of a fish across the gills and mucous membranes of the mouth. This would dilute the blood if it were not filtered off by the kidneys in a copious stream of very dilute urine. In salt water, on the other hand, there is a tendency for water to pass *out* from the body. To counteract this, marine fishes drink from the sea. This is shown by the fact that, if they are prevented from swallowing by introducing a rubber balloon into the oesophagus and inflating it, they soon lose so much water that they die. The fish kidney does not seem to have evolved the power of secreting a very concentrated urine so the problem remains of eliminating unwanted salt taken in with the sea water that is drunk. This is excreted, not by the kidney, but by special 'chloride excretory' cells situated in the gills. In migratory fishes, such as the salmon and eel, both types of regulatory mechanism are present, and are brought into action as required.

The Sea Lamprey (*Petromyzon marinus*) is another example of an anadromous fish that ascends the rivers of Europe in spring or early summer. Lampreys are parasitic, and not infrequently facilitate their migrations by attaching themselves with their sucker-like mouths to Striped Bass (*Roccus saxatilis*), salmon and other large fish bound in the same direction. They feed principally by sucking in the blood which flows from the wound, or liquidized flesh. Thus, they obtain not only free transport but also a meal *en route* for their spawning grounds. Here the two sexes undergo considerable changes in

Below: Pacific salmon leaping up a waterfall on their way to inland breeding grounds; right: their migratory routes to the North American waterways

Migration routes
of Pacific salmon

appearance. The males are often the first to arrive and at once begin nest-building; but they are soon joined by their mates. After spawning, the parents are so exhausted that none survive, but the young fry migrate until their own turn comes to reproduce.

Plaice (*Pleuronectes platersa*), flounder (*Platichthys flesus*), sole (*Solea solea*), smelts (*Osmerus* spp.), sturgeon and shad are other anadromous species that migrate to brackish or fresh water at the breeding season. The extent to which they may do so, varies, however, in different localities. The garfish (*Belone belone*) also migrates into very shallow, even brackish, coastal waters, where the young hatch and grow.

There are comparatively few catadromous fishes. Of these, the eel is undoubtedly the best known. As early as the fourth century BC, Aristotle was noting that: 'Some fishes leave the sea to go to the pools and rivers. The eel, on the contrary, leaves them to go down to the sea.' Aristotle's observations on eel migrations were repeated by Pliny and subsequent writers, but it was not until 1684 that the theory of an offshore breeding ground was proposed by the Tuscan author and poet, Francesco Redi, in the third volume of his book *Degli Animali Viventi negli Animali Viventi*. This book (*Animals Living in Living Animals*) was one of the first treatises on parasitology.

Subsequently it was realized that the fish-like animal, known as 'leptocephalus'—meaning thin head—was, in fact, the larval form of the eel. During the years 1904–22, the work on the Danish ocean-research vessels *Thor, Margrethe*, and *Dana* established that the spawning ground of the European Common Eel (*Anguilla anguilla*) is located in the Sargasso Sea. From here, the leptocephali are carried to Europe by the Gulf Stream, a journey lasting over $2\frac{1}{2}$ years. When it hatches, the leptocephalus is less than six millimetres ($\frac{1}{4}$ inch) in length; but it grows to be more than 75 millimetres (3 inches) before it reaches its destination. Bilaterally flattened, and leaf-like in shape, leptocephali are quite transparent, despite the presence of more than 100 vertebrae. No doubt this invisibility enables them to elude many a hungry predator during their hazardous journey.

On reaching the Continental Shelf of Europe, metamorphosis takes place and a typical cylindrical shape is acquired. This transition was first discovered from observations of many specimens brought to the surface by whirlpools in the Straits of Messina. 'Elvers', as the young eels are now called, ascend the rivers in countless numbers. They are instinctively obliged to adjust their bodies so that the water presses equally on both sides, and this keeps them on a direct course upstream. As they wriggle along, the bottom provides them with a point of reference relative to the flow of the stream. If they were moving with the current, and without some visual or tactile information about the movement of the

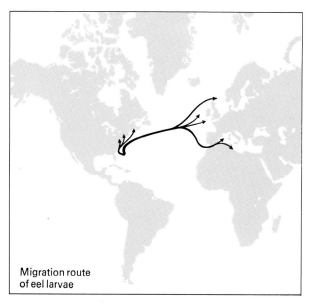

Migration route of eel larvae

Left: migration route of eel larvae from the Sargasso Sea; below: elvers of the European common eel swim upstream after their long larval migration

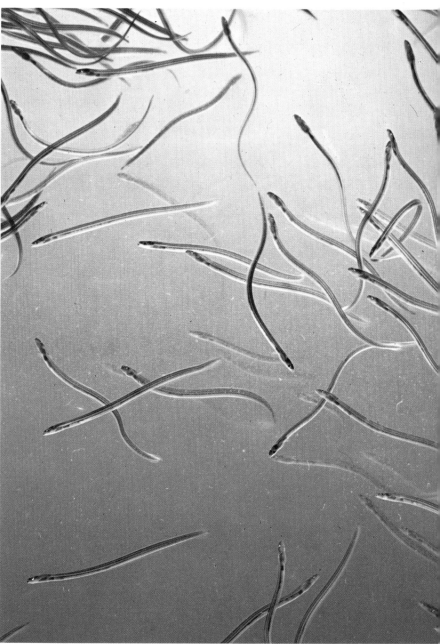

104

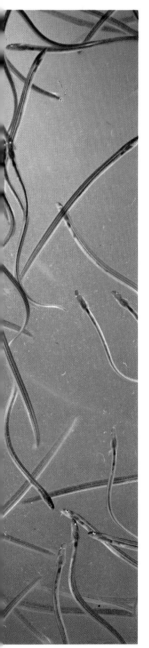

water in relation to the land, they would, of course, be unable to react to the current in any way. Elvers of American Eels have been reported in enormous numbers at the foot of the Niagara Falls, striving vainly to surround this impossable obstacle yet unable to swim in any direction other than upstream.

Eels remain in fresh water, feeding and growing, for several years. The differences between the sexes becomes apparent only when they are about six years old and the females begin to grow more strongly than the males. The males assume breeding-dress on reaching a length of 30–50 centimetres (12–20 inches)—that is, after $5\frac{1}{2}$–$6\frac{1}{2}$ years in fresh water. Females do so after $6\frac{1}{2}$–$8\frac{1}{2}$ years, by which time they have grown to an even larger size. Eels about a metre in length have probably lived for ten years or more since the elver stage. Sooner or later, however, the reproductive instinct overtakes them. They become silver in colour, their eyes enlarge, and they set out on a return journey to the Sargasso Sea.

The currently accepted theory is that the mature European Eels travel 5,600 kilometres (3,500 miles) back to their breeding grounds where they presumably spawn and die, for none return. It is a species distinct from the American Eel (*Anguilla rostrata*) which spawns earlier, and in a slightly different area. An alternative hypothesis, on the other hand, proposes that all migration attempts by European Eels are unsuccessful, and that the European population is of American origin, being maintained by American Eels which are successful in their shorter migration. According to this hypothesis, the two populations must belong to the same species and the differences between them are determined entirely by environmental factors.

Neither of these theories is entirely satisfactory, however, and a combination of ideas from both may provide the answer to the problem. The breeding ground in the Sargasso Sea may be common to eels of both American and European origin. The larvae that hatch there are distributed by ocean currents, both to America and to Europe, and the varying environmental conditions encountered along their routes determine the differences between the adults—European Eels have 110–119 vertebrae while American Eels have 103–111 vertebrae. Although it is reasonable to suppose that the American Eel would be capable of producing sufficient progeny to colonize Europe, North Africa and the east coast of America, it seems rather unlikely that all but the American Eels should be unable to reproduce successfully. Moreover, tagged eels removed from localities in European rivers, have been known to return to their homes after being released from distances of over 240 kilometres (150 miles). This indication of both navigational and long-distance travelling ability provides circumstantial evidence that they may migrate successfully to the Sargasso Sea.

Reptiles

The ancestors of marine reptiles were terrestrial animals which, at some stage during their evolutionary history, returned to the sea. Some of them became completely independent of land, but a fundamental aspect of the behaviour of turtles is that, although they may dwell in the ocean, they still nest on land. Only viviparous reptiles—those which give birth to living young—such as some sea-snakes, are able to reproduce in the water.

Certain sea-snakes periodically band together in the ocean, presumably for reproductive purposes. There is an old report of a migrating mass of millions of sea-snakes, twisted thickly together to form a procession three metres (10 feet) wide, extending for about 95 kilometres (60 miles); and several smaller aggregations have been seen more recently. Since most sea-snakes are viviparous, these gatherings are possibly no more than a means by which the two sexes come together at mating-time. Nevertheless, it seems unlikely that marine reptiles simply wander the ocean at random. Little is known of the movements of sea-going crocodiles, of which mass emigrations have been reported, but they are clearly restricted to specific geographical ranges and only inadvertently stray beyond them. The same was probably true of the extinct ichthyosaurs and plesiosaurs, which must have followed shoals of migrating fishes, yet remained within favourable climatic zones.

The sea-snakes that do not reproduce in water migrate to the land to give birth and rear their young. They make their nurseries in hollows of rocks in low islands and remain on land until the young have grown quite large.

Among marine turtles, there seems to be a worldwide pattern of migration between feeding territories and nesting beaches. This is necessary because the seaweeds on which turtles feed, grow expansively only in protected water, whereas beaches suitable for nesting occur where the waves break unhindered against the shore. These two types of habitat are seldom found close together, and in many cases, turtles have to travel many hundreds of kilometres from one to the other, returning to the same nesting beach year after year. A Hawkesbill Turtle (*Eretmochelys imbricata*) ringed by a Dutch officer in 1794 was found on the same Atlantic beach 30 years later.

The Green Turtle (*Chelonia mydas*) feeds on pastures of seaweed which may lie hundreds of kilometres out to sea. Green Turtles that feed off the coast of Brazil, annually migrate 2,250 kilometres (1,400 miles) to Ascension Island in the South Atlantic, where they breed. Their ability to navigate was known to seventeenth-century buccaneers, but how they find this tiny dot in the ocean, after battling so far against contrary winds and currents, is still not understood. Similarly, Atlantic Ridley Turtles (*Lepidochelys kempi*) gather from great

distances at their nesting places on the northern gulf coast of Mexico. Shortly after hatching, the baby turtles move directly to the sea which, of course, they have never seen before. They can do this by night or by day, and in all kinds of weather. When they are blindfolded, however, they get lost, so it would seem that they must, in some way, use their sense of sight.

Each spring the Leathery Turtle (*Dermochelys coriacea*) migrates to breed. Some go to the Bahamas, some to the Tortugas and some to the coast of Brazil. The Snapping Turtle (*Chelydra serpentina*), which lives mostly in fresh water, is said to make considerable journeys overland, but there is no evidence that these journeys are concerned with breeding. On the other hand, the Arrau (*Podocnemis expansa*), a fresh-water turtle of tropical South America, used to migrate in vast numbers to the Orinoco and Amazon, before its populations were drastically reduced by man. During the wet season, these turtles inhabit pools in the rain forest, but when the water subsides, they move to the rivers and migrate upstream and downstream to the chosen islands and sandbanks where they lay their eggs. This mass movement is called *arribaçaõ das tartarugas*, the ascent of the turtles. Unfortunately, it renders the creatures extremely vulnerable to local fishermen who have caught them for food since before the time of Columbus.

Penguins

Unable to fly because they have no normal wings—only flippers—penguins have to migrate by swimming. Some other sea-birds, which are good fliers, also migrate in this fashion. Brent Geese (*Branta bernicla*) and Eider Ducks (*Somateria mollissima*), for example, often swim for considerable distances during their winter dispersal, while Great Crested Grebes (*Podiceps cristatus*), and Red-necked Grebes (*P. grisegena*) follow the Swedish coast at about 45–200 metres (50–220 yards) from the shore throughout the day, feeding and moving in the same direction as that in which they fly at night.

The journeys of penguins are far more extensive. Species that nest on sub-Antarctic islands with a mild climate, such as the Falklands, Macquarie and Kerguelen, disperse through nearby waters after the breeding season, while Gentoo Penguins (*Pygoscelis papua*) spend the southern winters on the high seas and nest in South Georgia or South America. The greatest distances travelled are by Emperor Penguins (*Aptenodytes forsteri*) which, paradoxically, nest on the sea ice off Adelie Land during the southern winter and migrate northwards in spring. The solitary eggs are laid in places where the sea is often below freezing but kept open by currents.

At the end of the terrible polar winter, the birds migrate northwards—sometimes as far as the coast of Chile. Unmated individuals go first, followed by

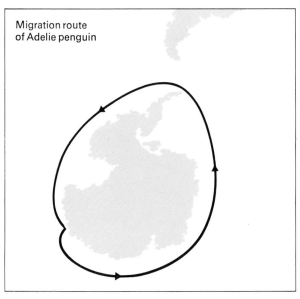

Migration route
of Adelie penguin

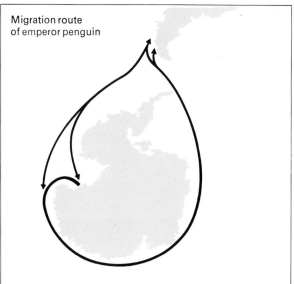

Migration route
of emperor penguin

The Adelie penguin migrates north and adopts an oceanic life after breeding in Antarctica during the summer

Emperor penguins breed in the same parts of Antarctica as the Adelie but do so during the winter, migrating north in the spring

Right: Jackass penguins head for their breeding grounds on the west coast of southern Africa

parents which have reared their single chicks. In addition to swimming, they walk for many kilometres across the ice and are transported on floes by the ocean currents. As winter again sets in, they return south chiefly underwater, moult on the border ice, and prepare to breed again as winter returns. Analysis of small pebbles taken from the stomachs of Emperor Penguins captured in the Bay of Whales, Antarctica, reveals that some of them were formed of kenyte, an igneous rock found only in the Ross Archipelago, 640 kilometres (400 miles) away.

In contrast, the Adelie Penguin (*Pygoscelis adeliae*) breeds in the same parts of Antarctica as the Emperor Penguin, but during the summer. It migrates to the north before winter sets in, adopting a pelagic life throughout much of the year. The Erect-crested Penguin (*Eudyptes sclateri*) breeds in large colonies, chiefly on Bounty Island and the Antipodes, then travels north to winter off the coasts of New Zealand. On the other hand, several species breed on the shores of New Zealand and South Africa, but are sedentary and never travel far from their birthplace.

Whales and seals

Although the tropics conjure up an image of luxurious vegetation and colourful coral reefs, the blue waters of the tropical seas are generally deficient in nutrients and consequently underpopulated. A dearth of plankton results in poor fisheries and comparatively few of the giant whalebone whales that feed on krill are found. Exceptions are the sea off the Galapagos Islands, the Caribbean Sea and the Gulf of Aden, where the waters teem with plankton, an abundance of fishes, schools of porpoises and dolphins, together with Sei Whales (*Balaenoptera borealis*), Bryde's Whales (*B. brydei*) and Sperm Whales (*Physeter catodon*), as well as migratory Blue Whales (*Balaenoptera musculus*), Fin Whales (*B. physalus*) and Humpbacks (*Megaptera nodosa*). These have to travel through thousands of kilometres of barren water from the polar seas, before reaching their tropical feeding grounds. While actually migrating, they find little food and, for at least four months of the year, their cavernous stomachs are almost empty. Yet, during this time, they may travel more than 6,400 kilometres (4,000 miles), some of them while pregnant or suckling their young.

The oldest report on the migration of Blue Whales dates back to the last century, when individuals caught at Norwegian whaling stations were found to carry within their bodies the fragments of American bomb-lances. This indicates that Blue Whales migrating north along the American coast must cross the North Atlantic. In 1954, the stomach of a whale killed off New Zealand was found to contain a tin of tooth-powder. Inside this was a piece of paper on which was written the name and address of a member of the crew of the *Willem Barendsz*. The tin had been thrown overboard, south of Madagascar, during the previous whaling season.

The systematic marking of whales was begun

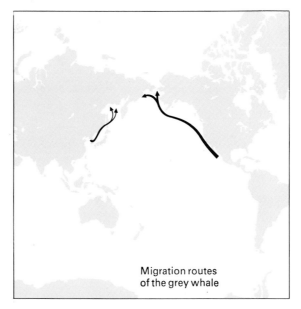

Migration routes
of the grey whale

about 1920, but it was not until at least 10 years later that whales began to be marked regularly by means of tubular metal cylinders fired at close range (not more than 20 metres, 65 feet) from a specially modified harpoon gun. The whales show no signs of reacting to these attentions, which are probably no more than pin-pricks to them. From the recovery of such marks, when whales have been subsequently caught, it is now known that Blue Whales and Fin Whales migrate each winter to tropical regions rich in food, such as the north-west coast of Africa, the Gulf of Aden and Bay of Bengal, returning year after year in summer to the same areas of the Antarctic Ocean.

Their routes do not take them so close to the mainland as do those of the Humpbacks, for the migrations of the three species do not coincide. Some sub-tropical whaling stations in South Africa and South America do, however, capture Fin Whales,

Left: migration routes of two populations of the grey whale; below: southern right whale showing baleen, the inner edges of which form a filter to retain the small animals on which the whale feeds

but mainly young animals are taken, and hardly any are caught from stations in the tropics themselves. Blue Whales are often seen off Tristan da Cunha during the southern winter, and here the Southern Right Whales (*Eubalaena australis*) give birth to their young. There are indications that pregnant Fin Whales may stay longer in the Antarctic than pregnant Blue Whales, and give birth to their young a few weeks later, but the winter quarters and migratory routes of Fin and Blue Whales are not yet precisely known, and not every individual migrates into warm waters during the winter season.

Although both Fin and Blue, as well as Humpback, Whales are known at times to cross the Equator, it is generally believed that southern populations are distinct from populations of the same species in the northern hemisphere. Members of both groups can be found in tropical waters every year, but they go there in different months and consequently do not intermingle. The Sei Whale is also cosmopolitan, migrating towards the poles in the spring, and towards the Equator in the autumn. Unlike some of the smaller whales, it keeps well clear of ice.

Sperm Whales move up the east coast of southern Africa in autumn, spend the winter in warmer waters where they calve and return to southern water in the Antarctic spring. Males which have been unable to secure a harem, or have not yet attained sexual maturity, follow them later in the year. Their staple food consists of giant squids which eat fishes that, in turn, depend on plankton. Dolphins of various species, which also feed on fishes, are known to be migratory; but other cetaceans keep to far more restricted areas. Greenland Whales (*Balaena mysticetus*), belugas (*Delphinapterus leucas*) and narwhals (*Monodon monoceros*) occur in Arctic waters, for instance, while Pigmy Right Whales (*Neobalaena marginata*) and Hourglass Dolphins (*Cephalorhynchus commersonii*) are restricted to the cold south where they have been seen over extensive areas. Porpoises (*Delphinidae*) and beaked whales (*Physeteridae*) are found only in the North Atlantic, although porpoises may migrate as far north as Spitzbergen and Greenland. They are nearly always seen near the coast, and are sometimes found quite a long way up the largest rivers.

Although sealions and seals are clumsy on land, they are skilful swimmers and divers. Some, such as the Harbour Seal (*Phoca vitulina*) remain on land most of the time, entering the water only to feed, but other species, including the Northern Fur Seal or sealion (*Callorhinus ursinus*), may spend up to eight months at sea. Most species are gregarious, but the Ross Seal (*Ommatophoia rossi*) lives quite alone in total darkness during the winter.

Whales give birth in the water, but seals, although they range over vast areas of ocean, return annually to the same restricted terrestrial breeding grounds.

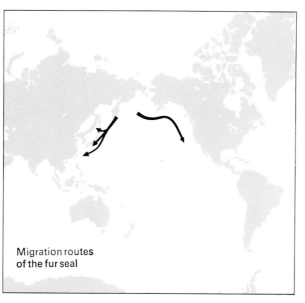

Migration routes of the fur seal

Left: *migration routes of two populations of northern fur seals; below: harbour seals resting on a rock ledge above the water. Harbour seals usually enter the water only to feed, but the northern fur seal has been known to spend as long as eight months at sea*

Colonies vary in size from a few individuals to more than a million animals crowded together on the same beach. At such times, unfortunately, the immature animals are especially vulnerable to cropping for their fine skins.

The most spectacular migrations are those of the Northern Fur Seals whose breeding grounds are in two groups of small islets, the Pribilov and Commander Islands. These are situated in the Bering sea between Siberia and Alaska, but the populations of seals that breed on them do not intermingle. At the end of the breeding season, seals from the Commander Islands swim south-west into the Pacific, sometimes as far as Japan, while those from the Pribilov Islands migrate south-eastwards to the coasts of California. The round trip in each case is nearly 10,000 km. (6,000 miles), yet the animals are able to navigate accurately each year back to the small islands on which they were born.

Alternatives
to Migration

Although migration is part of the instinctive life-cycle of many animals all over the world, it is, nevertheless, only one method of avoiding unfavourable conditions. There are many animals which do not migrate and which have evolved other ways of coping with inclement seasons. Dormancy is one of them.

Dormancy may be an immediate response to adverse conditions, in which case recovery occurs soon after these conditions have disappeared. For example, insects in a state of cold stupor resume activity when the temperature rises. On the other hand, many animals that hibernate in winter do so in the physiological resting state known as 'diapause' which is under hormonal control and from which recovery is not immediate.

Diapause, like navigation, depends upon the operation of the 'biological clock' and is a response to changing day-length or 'photoperiod'. In this way, it can be induced by environmental changes which take place before the advent of the adverse conditions to which dormancy is an adaptation.

The state of diapause is usually characterized by the temporary failure of growth or reproduction, by reduced metabolism, and often by enhanced resistance to climatic factors such as heat, cold, or drought, and by other morphological, physiological and behavioural characteristics. The phenomenon is widespread among living organisms, and is an especially striking feature of the biology of insects and mites.

The physiology of diapause has been studied intensively. A genetically determined state of suppressed development, which may be induced by photoperiod, diapause is an important adaptive mechanism for surviving not only periods of unfavourable environmental conditions, but also periods during which food is scarce or absent. Indeed, one of the simplest aspects of seasonal timing is concerned with food supply. The life-cycle of an animal must be synchronized with that of its natural food. For instance, some of the insects that feed on leaves cease feeding when these wither and turn brown in the fall. They or their larvae then burrow into the soil where they hibernate until the following spring. By the time that their period of diapause has ended, foliage is again available for them to feed on. Other insects pass the winter as diapausing eggs which hatch the following spring. The parent insects are stimulated by decreasing photoperiod to lay diapausing eggs that will not hatch until they have been exposed to frost. The significance of the photoperiodic response is that seasonal adjustments can be made before cold, drought or starvation puts an end to all activity.

Many insects show specific patterns of behaviour which are associated with the physiological preparations for diapause. Larval forms often seek out suitable hiding-places in the soil or under stones or bark, sometimes reversing their customary reactions to gravity and light. Other species construct special cocoons, in which they will be protected during their period of dormancy. Overwintering pupae of some species have concealing coloration designed to camouflage them in the absence of foliage.

Seasonal changes in appearance are often closely associated with diapause. For instance, the spring form of certain butterflies is produced from diapause pupae resulting from the short days of the previous autumn; while the summer form, whose coloration is quite different, emerges from non-diapausing pupae. By exposing caterpillars to the appropriate photoperiod, either form can be produced experimentally.

Long day-lengths lead to the production of female aphids whose eggs develop rapidly and without fertilization. Short day-lengths, on the other hand lead to the production of both males and females whose fertilized eggs pass the winter in a state of embryonic diapause. Similar relationships between photoperiod, reproduction and diapause have been reported in various insects, mites, water-fleas, and so on.

Reproductive rhythms are likewise important in determining seasonal peaks of numbers in populations of aquatic animals. Most species undergo periods of dormancy when the active stages disappear and the species is maintained by eggs, or by

*Facing page:
hibernating monarch
butterflies in Mexico
festooning the trees on
which they are resting*

developing larvae, resting in diapause. In the waters of temperate regions, seasonal changes of form are also conspicuous among various species of plankton. During the summer, while rapid multiplication is taking place, there may be a temporary predominance of varieties with well-marked crests, elongated spines and other extensions of the body surface, compared with the more compact forms of the winter generation. These variations may be adapted to seasonal changes in the temperature of the water, which affects its density and viscosity. Reduced viscosity in summer increases the difficulty of floating and is compensated for by the crests and spines. This type of seasonal change in form is limited to shallow lakes with a wide annual range of temperature.

Many larger animals show seasonal breeding cycles, and reproduction is arrested throughout the remainder of the year. Not only is breeding usually controlled by photoperiod—except in amphibians and some other cold-blooded forms that respond directly to warmth—but reproduction frequently alternates with migration so that unfavourable seasons are avoided. Reptiles and most of the mammals that do not migrate, frequently spend the winter in hibernation, a condition brought about mainly in response to low temperature, although day-length also has an effect.

Mammalian hibernation differs from insect diapause in that it is interrupted by short, periodic arousals throughout the winter. It enables animals to survive the winter without feeding and with minimum expenditure of food reserves on metabolism. Temperature is regulated at about 1°C above the ambient in the range 5°C–15°C (41°F–59°F), and arousal can be induced by temperatures above and below this range.

There have been few more extraordinary discoveries about birds in modern times than those which led, in some measure, to the validation of the discredited seventeenth-century belief that birds may, in certain circumstances, tide over periods of stress in a state of torpidity. During an extremely cold spell in January 1913, eight White-throated Spine-tailed Swifts (*Hirundapus caudacutus*) were taken out of a crevice in the cliffs of Slover Mountain, California, where they, with many others, were roosting in a dazed or numb state. During the nineteenth century torpidity had been observed in hummingbirds, but it was not until the chance discovery, in 1946, of typical hibernation in the poorwill (*Phalaenoptilus nuttallii*) that widespread interest in the phenomenon was aroused.

More recently, torpidity as an alternative to migration has been described in the Australian White-backed Swallow (*Cheramoeca leucosterna*), Crimson Chats (*Epthianura tricolor*), Banded White-face (*Aphelocephala nigricincta*), Red-capped Robin Flycatcher (*Petroica goodenovii*), White-fronted Honeyeater (*Melidectes leucostephes*) and Mistletoe

Bird (*Dicaeum hirundinaceum*). If it could suspend animation, a small bird could survive the cold winter nights of the Australian desert more economically than if it had to ingest extra food to keep warm. Such torpidity is probably a direct response to the cold and not determined by photoperiodism.

The road to nowhere

Migration must clearly be seen as only one of several alternative strategies by which different animals respond to the seasonal or irregular changes in their environments which affect their requirements of food, reproduction and space in which to live. Nevertheless the migratory instinct has been implanted in animals of many and varied types. At the same time, the progress of evolution results from innumerable compromises, and migration does have drawbacks as well as advantages. In particular, migrating animals, far from the security of their permanent homes, are particularly vulnerable to the unwelcome attentions of enemies, especially man.

Because they are bunched together in their migratory flyways, birds can be subjected to intensive slaughter. The Passenger Pigeon of North America is a case in point. At one time, this species existed in greater numbers than any other land-bird, and ranged from the Atlantic Ocean to the Rocky Mountains and from Northern Canada to Mexico. The total population has been estimated at over 3,000 million. In comparison, the total number of land-birds in the British Isles today is probably not more than 200 million.

Before the settlers came to America, Passenger Pigeons used to migrate in massed flights of millions of birds in tightly packed ranks. They were responding to fluctuations in the annual crop of their favourite foods—acorns, beech-nuts and chestnuts. But they destroyed crops and themselves provided good food, so they were shot and trapped in great numbers. So dense were the migrating masses that over 100 birds could be killed with a single shot. They were even knocked down with long poles, and 1,680 pigeons were once captured in the single throw of a net. Large flocks could still be seen in 1880 but, by the end of the century, Passenger Pigeons were virtually extinct in the wild. The last caged specimen died in captivity in the Cincinnati Zoo in 1914. The bird fauna of Europe is likewise impoverished by the massacre of migrants as they pass through Italy and other southern countries.

Wild animals are no respecters of international frontiers, and however thoroughly they may be protected in one part of their range, migratory species may well be endangered in another. Nor is the problem merely one of restricting the activities of sportsmen and hunters. The finely balanced process of migration can be upset when habitats are changed or destroyed by agriculture, forestry and drainage schemes.

African elephants do considerable damage to trees when present in large numbers and prevented by human settlement from migrating as they did in the past

Migratory game animals of the African savanna may be extremely vulnerable. Elephants, for instance, are relatively safe in National Parks so long as they can be protected from poachers, but once they stray beyond the borders, every man's hand is against them. Not only are they extremely destructive to farms and plantations, but they provide valuable meat and ivory. By blocking migration routes, human settlement has largely restricted these animals to the National Parks where over-grazing and soil erosion frequently occur.

There are two schools of thought regarding the management of big game, especially elephants. According to one, the population increase should be limited by controlled shooting. It is considered best to eliminate entire herds rather than to select individual animals for slaughter, because this does not upset family groupings and cause unnecessary suffering. Some biologists, however, feel that cropping runs counter to the ethics of game conservation. They believe that the present population explosion among the elephants of East Africa is part of a normal cycle in numbers which will be controlled by natural processes. In support of this they cite the curious behaviour of elephants in Tsavo National Park where droughts normally occur about every 10 years. Around 1956, a year in which there was no drought, the elephants began to destroy the commiphora (myrrh) trees and they

continued to do so the following year. During the severe drought of 1960–61, 300 rhinos starved to death along a 64-kilometre (40-mile) stretch of the Athai River; and it was believed that the previous effects of destruction by elephants of the trees would result in desert conditions everywhere. But, when the rains returned, the commiphora trees were replaced by seedling acacias, and the devastated country produced a greater abundance of perennial grasses than ever before. These were cropped by the elephants and other game and the country was able to support a greater concentration of wildlife than previously.

Nomadism, too, has its drawbacks. Although ungulates are better developed and more mobile at birth than the blind, naked offspring of many other mammals, the young of most species remain hidden and largely immobile for some days or even weeks after birth. In probably not more than 40 out of 185 kinds, do the young follow their mothers from the time they are able to stand. The wildebeest is one of four or five species of antelopes known to have young that follow their mothers: the young of other antelopes are hiders. Each of the two strategies includes a number of adaptations that enhance its efficiency. The measures that promote concealment are remarkably similar, even among distantly related species, whereas there are several different strategies for safeguarding follower young, which

are apt to be more exposed to predation even though they are less helpless than are hiders. In some species, such as the rhinoceros, hippo and giraffe, the mother is able to defend her young against lions and Spotted Hyenas (*Crocuta crocuta*) in packs. Musk-oxen form defensive rings around their calves, but the young of wildebeest, topi (*Damaliscus korrigum*), blesbok and caribou have to escape from their predators by flight.

The species whose young follow their mothers and depend upon flight, inhabit open country where they enjoy a nomadic or migratory existence. In the case of the wildebeest and other ungulates in the same category, there are two other characteristics: a brief, synchronized peak of reproduction and the formation of large aggregations. Even at the Equator, where an extended period of breeding is usual, most wildebeest are born during an interval of about three weeks. Survival is greater among calves born in aggregations than it is in smaller herds. Lack of reproductive synchrony between small herds inhabiting different localities renders their calves more vulnerable to predation by Spotted Hyenas than calves born to wildebeest in large aggregations. Furthermore, there are enough older calves even in the larger aggregations, to make it difficult for Spotted Hyenas to single out the younger and more vulnerable individuals. Reproductive synchrony—whereby populations of predators are temporarily glutted without being sustained with a regular supply of food over an extended period and the formation of large aggregations—whereby the most vulnerable calves are protected by the confusion created by large numbers—are the keys to successful breeding in the wildebeest.

Emigration acts like a relief valve to the pressure of excess population. In itself, it usually only benefits a species indirectly. Unless all is well with the natural world, however, the population increases that result in emigration seldom occur. For this reason, it seems unlikely that a mass irruption of South Africa's migratory springbuck will ever again be witnessed.

One becomes so accustomed to the ruthless efficiency of natural selection, and to the marked economy of natural processes, that examples of apparent waste cannot be accepted without careful assessment. The possible ecological significance of mass emigration has already been discussed, but another consideration is the surplus killing by carnivores which, under certain circumstances, may slaughter many more prey animals than they can possibly consume at one time. The fox in the hen-house provides a spectacular example. Other examples now known include the occasional massacre by foxes of Black-headed Gulls (*Larus ridibundus*) and Sandwich Terns (*Sterna sandvicensis*) in breeding colonies. In one instance, up to 230 birds were killed by each fox, but fewer than three per cent of the victims were eaten. The forays of the foxes occurred only on nights too dark for flight, and tracks in the sand near the carcasses indicated that the birds had made little or no attempt to escape.

Poor visibility also seems to play some part in the occasional mass slaughter of Thomson's Gazelle (*Gazella thomsoni*) in Serengeti, where 82 gazelles were found dead on one occasion. Only a few of the dead animals had been eaten and hyena tracks near the bodies gave no indication of a chase. Narwhals (*Monodon monoceros*), too, have been killed in large numbers when trapped by ice in small pools so that they were easy prey for Polar Bears (*Thalarctos maritimus*). Unnecessary slaughter seems not only wasteful, but against the killer's own interests. Not only does it deplete potential food supplies, but it wastes the predator's energy, even though it may provide useful practice in the art of killing.

It seems most probable, therefore, that unnecessary killing may result as much from the failure of the prey to escape from the predator as from the instinct of the latter to kill. The carnivore's hunting behaviour normally allows it to make the best use of its food resources, but there is nothing to prevent a sated predator from killing prey that is unable to escape. Consequently, surplus killing should be viewed, not as the result of mere maladaptations, but as a necessary and relatively small disadvantage accompanying generally efficient behavioural compromises. In general, it would not pay a gazelle to dash about on pitch-dark nights; therefore the apparent lack of response may, on the whole, have a selective advantage, even if on occasions its disadvantages show spectacularly. At the same time, to prevent surplus kills from occurring, carnivores would have to possess a special inhibition of hunting and killing—not after satiation, for this they already possess, but after receipt of information that further carcasses could not be utilized.

Emigration, too, is a phenomenon the disadvantages of which may be spectacular but which must, in the long run, have a selective advantage that more than outweighs them. This type of reasoning becomes an article of faith to the biologist.

Even if the tentative explanation of the advantages of emigration, given earlier, should turn out to be incorrect, we may, nevertheless, remain convinced that some distinct advantage must accrue from migration in all its aspects, otherwise the phenomenon would never have evolved. This advantage makes migration a dominant, widespread instinct, but if, at any time, the balance should be tilted so that it became generally disadvantageous to a species, we may be equally sure that the extinction of that species would inevitably follow. That is why species adapted so that migration is an essential part of their life-cycle are particularly vulnerable to harmful human activities. We have only to alter the balance so that they no longer benefit from their journeys for them to be endangered. This is

Facing page: the effect of overgrazing at Tsavo, resulting from a population explosion among elephants unable to emigrate from the region

116

what happened to the Passenger Pigeon. Even the exhausting trans-Saharan migration must still be well worthwhile to the birds that undertake it, even though it probably evolved before the desert was as extensive as it is today. If it were no longer profitable, cuckoos, swifts and swallows, among others, could soon become extinct.

There are occasions when migratory behaviour may extend dramatically the geographical range of a species. An example of this is afforded by the Cattle Egret (*Bubulcus ibis*) of the Mediterranean, Near East and Africa, which has come to the New World—presumably without the aid of man. The first observations of this species in the Western Hemisphere were made in Guiana towards the end of the last century, and it was present in Columbia as early as 1917. Since that time, Cattle Egrets have spread and become abundant throughout much of northern South America, Central America, the Caribbean Islands, United States and Southern Canada.

Another example is the Brown Rat (*Rattus norvegicus*) which was unknown in western Europe until the eighteenth century. In 1727, however, great hordes were seen marching westwards from Russia, and swimming across the Volga. Later the whole of Europe was occupied by Brown Rats, and within two centuries, Black Rats (*Rattus rattus*) were largely replaced by them. Today, Black Rats are found chiefly on ships and in ports. The Brown Rat tends to live in sewers and outhouses, and its habits do not bring it into close contact with man. Consequently, since that time, bubonic plague, which had been largely transmitted from Black Rats by their fleas, ceased to be a serious epidemic disease of man in Europe.

A species can only establish itself in a new environment if there is a vacant space or ecological niche available to it in the ecosystem, or if it is more successful than another species, already established, which it is able to supplant. In different regions of the world, there are comparable niches which may be occupied by quite unrelated species. Thus, the numerous species of antelopes on the African savannas are paralleled by a diversity of marsupial kangaroos and wallabies in Australia. When rabbits were introduced by man, they multiplied partly at the expense of the native marsupial fauna of the Australian plains, but in an almost vacant niche. Cattle Egrets likewise seem to have found a vacant niche in the New World, which they were able to exploit satisfactorily.

In the immense field of research concerned with migration and navigation, every new discovery raises new hypotheses and further questions. Only decades ago, both these phenomena were shrouded in mystery, and it seemed almost as though some form of extra-sensory perception might be involved. In contrast, the hypotheses of today are firmly based on observation and experiment. Although much still remains to be discovered, we know enough to provide a general outline of animal migration, its timing, and the means of navigation employed. We are also aware of the risks to which migratory species are subjected. There is no excuse for not applying this knowledge wisely and well.

Bibliography

CARTHY, J. D. *Animal Navigation* (Allen & Unwin) 1956.

CLOUDSLEY-THOMPSON, J. L. *Rhythmic Activity in Animal Physiology and Behaviour* (Academic Press) 1961.

CLOUDSLEY-THOMPSON, J. L. *Animal Twilight, Man and Game in Eastern Africa* (Foulis) 1967.

DORST, J. *The Migrations of Birds* (Heinemann) 1962.

DORST, J. *The Life of Birds: Vol. II* (Weidenfeld & Nicolson) 1974.

GRIFFIN, D. R. *Bird Migration* (Heinemann) 1965.

HARDEN JONES, F. R. *Fish Migration* (Arnold) 1968.

HEAPE, W. *Emigration, Migration and Nomadism* (Heffer) 1931.

JOHNSON, C. G. *Migration and Dispersal of Insects by Flight* (Methuen) 1969.

LOCKLEY, R. M. *Animal Migration* (Arthur Barker) 1967.

MATTHEWS, G. V. D. *Bird Navigation* (Collins, New Naturalist Series) 1948.

SCHMIDT-KOENIG, K. *Migration and Homing in Animals* (Springer-Verlag, Berlin) 1975.

THOMPSON, A. L. *Bird Migration* (Witherby) 1936.

URQUHART, F. A. *The Monarch Butterfly* (Oxford University Press) 1960.

WILLIAMS, C. B. *The Migration of Butterflies* (Edinburgh) 1930.

WILLIAMS, C. B. *Insect Migration* (Collins, New Naturalist Series) 1958.

Acknowledgements

Half-title page: Jacana; Title Page G.D. Lepp/Brower; Lepp Associates; Contents page: Bruce Coleman; 6 Tollu/Jacana; 7 R. Maltini–P. Solaini; 9 top IGDA; 8–9 Peter Ward; 12 Di Giuliano Cappelli; 13 Françis Roux/Jacana; 14 Norman Myers/Ardea; 16–17 Pierre Petit/Jacana; 18 Ardea; 19 Varin-Visage/Jacana; 20–21 Des Bartlett/Bruce Coleman; 22 IGDA; 22–23 NHPA; 24 top Peter Ward, bottom A. Kerneis/Jacana; 25 S. C. Bisserot/Bruce Coleman; 26–27 J. Dubois/Jacana; 28 A. Margiocco; 30 Maes/Jacana; 31 top Richard Vaughan, 31 bottom G. Cappelli/Borgioli; 32 bottom left John Sundance/Jacana; 32–33 Visage-Varin/Jacana; 33 Ziesler/Jacana; 34 Eric Hosking; 35 top Pierre Petit/Jacana, 35 bottom Albert Visage/Jacana; 36 Ziesler/Jacana; 37 top Bruce Coleman, bottom J. Robert/Jacana; 38 top R. Thibout/Jacana, bottom J. Robert/Jacana; 39 Marka; 40 Claude Nardin/Jacana; 40–41 M. Krishnan/Ardea; 41 A. Kerneis/Jacana; 42 top Joe van Wormer/Bruce Coleman, centre Varin-Visage/Jacana; 42–43 André Ducrot/Jacana; 44 G. Haüsle/Jacana; 45 Varin-Visage/Jacana; 46–47 J. Dubois/Jacana; 47 André Ducrot/Jacana; 48 P. A. Milwaukee/Jacana; 49 top Donald D. Burgess/Ardea, 49 bottom Rieuse/Jacana; 50–51 Brian Hawkes/Jacana; 52 Jacana; 52–53 Tollu/Jacana; 53 Francisco Grize/Bruce Coleman; 54 P. Morris/Ardea; 55 Jane Burton/Bruce Coleman; 56 S. C. Bisserot/Bruce Coleman; 57 I. R. Beames/Ardea; 58 John Markham/Bruce Coleman; 59 Peter Ward; 60 Jane Burton/Bruce Coleman; 61 top D. Houston/Bruce Coleman, bottom Ardea; 62–63 L. Lacosta/Jacana; 64 G. D. Lepp/Brower-Lepp Associates; 65 C. & M. Moiton/Jacana; 66 C. & M. Moiton/Jacana; 67 Jacana; 68 P. Lorne/Jacana; 69 Jacana; 70–71 John Mason/Ardea; 71 S. C. Bisserot/Bruce Coleman; 72 C. J. Ott/Bruce Coleman; 73 Chuck Feil/Marka; 74 Cecil Rhode/Frank Lane; 74–75 Clem Haagner/Ardea; 75 Marka; 76–77 S. Gillsater/Bruce Coleman; 78–79 Massart/Jacana; 78 Steve McCutcheon/Marka; 79 A. Rainon/Jacana; 80 Chuck Feil/Marka; 80–81 Bert Leidmann/Zefa; 82 Chuck Feil/Globe; 83 Philcarol/Frank Lane; 84 Bob Campbell/Bruce Coleman; 84–85 J. Robert/Jacana; 86–87 Clem Haagner/Ardea; 87 Jacana; 88 Steve McCutcheon/Marka; 89 W. E. Ruth/Bruce Coleman; 90–91 Spectrum; 91 Peter Ward; 92 Jane Burton/Bruce Coleman; 93 top I. & L. Beames/Ardea; 93 Hans Reinhard/Bruce Coleman; 94 Jacana; 95 Bruce Coleman; 96 Soulaine-Cedri; 97 C. Carre/Jacana; 98 C. Carre/Jacana; 99 Allan Power/Bruce Coleman; 100 Horace Dobbs/Bruce Coleman; 100–101 D. P. Wilson; 101 Ron & Valerie Taylor/Ardea; 102–103 R. Thompson; 104–105 Jane Burton/Bruce Coleman; 106 Peter Ward/Popperfoto; 107 Fabius Henrion/Jacana; 108–109 Eliott/Jacana; 110 Jen & Des Bartlett/Bruce Coleman; 111 William E. Ferguson; 112 G. D. Lepp/Brower-Lepp Associates; 114–115 Alan Weaving/Ardea; 117 Heather Angel.

Index

119